STUDENT IN-CLASS NOTEBOOK

CHARLES A. DANA CENTER
The University of Texas at Austin

MYMATHLAB® FOR FOUNDATIONS OF MATHEMATICAL REASONING

Charles A. Dana Center
The University of Texas at Austin

D1469194

PEARSON

Boston   Columbus   Indianapolis   New York   San Francisco
Amsterdam   Cape Town   Dubai   London   Madrid   Milan   Munich   Paris   Montreal   Toronto
Delhi   Mexico City   São Paulo   Sydney   Hong Kong   Seoul   Singapore   Taipei   Tokyo

About the Charles A. Dana Center at The University of Texas at Austin

The Dana Center develops and scales math and science education innovations to support educators, administrators, and policy makers in creating seamless transitions throughout the K–14 system for all students, especially those who have historically been underserved.

We work with our nation's education systems to ensure that every student leaves school prepared for success in postsecondary education and the contemporary workplace—and for active participation in our modern democracy. We are committed to ensuring that the accident of where a student attends school does not limit the academic opportunities he or she can pursue. Thus, we advocate for high academic standards, and we collaborate with local partners to build the capacity of education systems to ensure that all students can master the content described in these standards.

Our portfolio of initiatives, grounded in research and two decades of experience, centers on mathematics and science education from prekindergarten through the early years of college. We focus in particular on strategies for improving student engagement, motivation, persistence, and achievement.

We help educators and education organizations adapt promising research to meet their local needs and develop innovative resources and systems that we implement through multiple channels, from the highly local and personal to the regional and national. We provide long-term technical assistance, collaborate with partners at all levels of the education system, and advise community colleges and states.

We have significant experience and expertise in the following:

- Developing and implementing standards and building the capacity of schools, districts, and systems
- Supporting education leadership, instructional coaching, and teaching
- Designing and developing instructional materials, assessments, curricula, and programs for bridging critical transitions
- Convening networks focused on policy, research, and practice

The Center was founded in 1991 at The University of Texas at Austin. Our staff members have expertise in leadership, literacy, research, program evaluation, mathematics and science education, policy and systemic reform, and services to high-need populations. We have worked with states and education systems throughout Texas and across the country. For more information about our programs and resources, see our homepage at **www.utdanacenter.org**.

About the New Mathways Project

The NMP is a systemic approach to improving student success and completion through implementation of processes, strategies, and structures based on four fundamental principles:

1. Multiple pathways with relevant and challenging mathematics content aligned to specific fields of study
2. Acceleration that allows students to complete a college-level math course more quickly than in the traditional developmental math sequence
3. Intentional use of strategies to help students develop skills as learners
4. Curriculum design and pedagogy based on proven practice

The Dana Center has developed curricular materials for three accelerated pathways—Statistical Reasoning, Quantitative Reasoning, and Reasoning with Functions I and II (a two course preparation for Calculus). The pathways are designed for students who have completed arithmetic or who are placed at a beginning algebra level. All three pathways have a common starting point—a developmental math course that helps students develop foundational skills and conceptual understanding in the context of college-level course material.

In the first term, we recommend that students also enroll in a learning frameworks course to help them acquire the strategies—and tenacity—necessary to succeed in college. These strategies include setting academic and career goals that will help them select the appropriate mathematics pathway.

In addition to the curricular materials, the Dana Center has developed tools and services to support project implementation. These tools and services include an implementation guide, data templates and planning tools for colleges, and training materials for faculty and staff.

Acknowledgments

The development of this course began with the formation of the NMP **Curricular Design Team**, who set the design standards for how the curricular materials for individual NMP courses would be designed. The team members are:

Richelle (Rikki) Blair, Lakeland Community College (Ohio)

Rob Farinelli, College of Southern Maryland (Maryland)

Amy Getz, Charles A. Dana Center (Texas)

Roxy Peck, California Polytechnic State University (California)

Sharon Sledge, San Jacinto College (Texas)

Paula Wilhite, North Texas Community College (Texas)

Linda Zientek, Sam Houston State University (Texas)

The Dana Center then convened faculty from each of the NMP codevelopment partner institutions to provide input on key usability features of the instructor supports in curricular materials and pertinent professional development needs. Special emphasis was placed on faculty who need the most support, such as new faculty and adjunct faculty. The **Usability Advisory Group** members are:

Ioana Agut, Brazosport College (Texas)

Eddie Bishop, Northwest Vista College (Texas)

Alma Brannan, Midland College (Texas)

Ivette Chuca, El Paso Community College (Texas)

Tom Connolly, Charles A. Dana Center (Texas)

Alison Garza, Temple College (Texas)

Colleen Hosking, Austin Community College (Texas)

Juan Ibarra, South Texas College (Texas)

Keturah Johnson, Lone Star College (Texas)

Julie Lewis, Kilgore College (Texas)

Joey Offer, Austin Community College (Texas)

Connie Richardson, Charles A. Dana Center (Texas)

Paula Talley, Temple College (Texas)

Paige Wood, Kilgore College (Texas)

Some of the content for this course is derived from the Quantway™ course, which was developed under a November 30, 2010, agreement by a team of faculty authors and reviewers contracted and managed by the Charles A. Dana Center at The University of Texas at Austin under sponsorship of the Carnegie Foundation for the Advancement of Teaching. Quantway™ is copyright © 2011 by the Carnegie Foundation for the Advancement of Teaching and the Charles A. Dana Center at The University of Texas at Austin. Statway™ and Quantway™ are trademarks of the Carnegie Foundation for the Advancement of Teaching.

Development of the *Foundations for Mathematical Reasoning* course was made possible by a grant from the Kresge Foundation. Additional funding and support for the New Mathways Project was provided by Carnegie Corporation of New York, Greater Texas Foundation, Houston Endowment, Texas legislative appropriations request, and TG.

Any opinions, findings, conclusions, or recommendations expressed in this material are those of the author(s) and do not necessarily reflect the views of these funders or The University of Texas at Austin. This publication was also supported through a collaboration between the Charles A. Dana Center, Texas Association of Community Colleges, and Pearson Education, Inc.

Project Lead and Authors

April Andreas, associate professor, engineering, McLennan Community College (Texas)

Pauline Chow, senior professor, mathematics, Harrisburg Area Community College (Pennsylvania)

Christina Hoffmaster, instructor, Vernon College (Texas)

Connie J. Richardson, advisory lead, Charles A. Dana Center

Francisco Savina, curriculum lead and lead author, Charles A. Dana Center

Randell Simpson, resident mathematics instructor, Temple College (Texas)

Charles A. Dana Center Project Staff

Adam Castillo, graduate research assistant

Heather Cook, project manager

Ophella C. Dano, lead production editor

Rachel Jenkins, consulting editor

Phil Swann, senior designer

Sarah Wenzel, administrative associate

Amy Winters, lead editor (freelance)

Math Faculty Reviewers from the NMP Codevelopment Teams

Alamo Colleges–Northwest Vista College, San Antonio, Texas

Austin Community College, Austin, Texas

Brazosport College, Lake Jackson, Texas

El Paso Community College, El Paso, Texas

Kilgore College, Kilgore, Texas

Lone Star College–Kingwood, Kingwood, Texas

Midland College, Midland, Texas

South Texas College, McAllen, Texas

Temple College, Temple, Texas

Quantway™ version 1.0 (2011)	*Foundations* version 1.0 (2013)
Stuart Boersma, professor of mathematics, Central Washington University (Washington)	Eileen Faulkenberry, associate professor of mathematics, Tarleton State University (Texas)
Mary Crawford-Mohat, associate professor in mathematics, Onondaga Community College (New York)	Christina Hoffmaster, Vernon College (Texas)
Margaret (Peg) Crider, professor in mathematics, retired, Lone Star College (Texas)	Tom Faulkenberry, assistant professor of mathematics, Tarleton State University (Texas)
Caren Diefenderfer, professor of mathematics, Hollins University (Virginia)	Liz Scott, San Augustine Independent School District (Texas)
Amy Getz, manager of community college services, Charles A. Dana Center, University of Texas at Austin (Texas)	Jack Rotman, Lansing Community College (Michigan)
Michael Goodroe, lecturer of mathematics and learning support liaison, Gainesville State College (Georgia)	Jeff Morford, Henry Ford Community College (Michigan)
Cinnamon Hillyard, assistant professor in mathematics, University of Washington Bothell (Washington)	Constance Elko, Austin Community College (Texas)
Robert Kimball, professor in mathematics, retired, Wake Technical Community College (North Carolina)	
Deann Leoni, professor of mathematics, Edmonds Community College (Washington)	
Michael Lundin, professor of mathematics, Central Washington University (Washington)	
Bernard L. Madison, professor in mathematical sciences, University of Arkansas (Arkansas)	
Jeffrey Morford, professor of mathematics, Henry Ford Community College (Michigan)	
Jane Muhich, managing director for community college program development, and director of productive persistence, Carnegie Foundation for the Advancement of Teaching (California)	
Julie Phelps, professor of mathematics, Valencia College (Florida)	

Pearson Education, Inc. Staff

Vice President, Editorial Jason Jordan
Strategic Account Manager Tanja Eise
Editor in Chief Michael Hirsch
Senior Acquisitions Editor Dawn Giovanniello
Editorial Assistant Megan Tripp
Digital Instructional Designer Tacha Gennarino
Manager, Instructional Design Sara Finnigan
Senior Project Manager Dana Toney
Director of Course Production, MyMathLab
  Ruth Berry
MathXL Content Developer Kristina Evans

Project Manager Kathleen A. Manley
Project Management Team Lead Christina Lepre
Product Marketing Manager Alicia Frankel
Senior Author Support/Technology Specialist
  Joe Vetere
Rights and Permissions Project Manager
  Gina Cheselka
Procurement Specialist Carol Melville
Associate Director of Design Andrea Nix
Program Design Lead Beth Paquin
Composition Dana Bettez

Contents

Lesson	Lesson Title and Description	

Student Resources

Curriculum Overview

Contents

- About *Foundations of Mathematical Reasoning*
- Structure of the curriculum
- Constructive perseverance level
- The role of the preview and practice assignments
- Resource materials
- Language and literacy skills
- Curriculum design standards
- Prerequisite assumptions
- Learning goals
- Content learning outcomes

About *Foundations of Mathematical Reasoning*

Foundations of Mathematical Reasoning is a semester-long, quantitative literacy-based course designed to provide you with the skills and conceptual understanding to succeed in a college-level statistics, quantitative literacy, or STEM path algebraic reasoning course.

Foundations of Mathematical Reasoning is organized around big mathematical and statistical ideas. The course will help you develop conceptual understanding and acquire multiple strategies for solving problems (as described in the NMP Curriculum Design Standards on page xvii). The course will prepare you for success in future courses and will help you develop skills for the workplace and as a productive citizen.

You are encouraged (or your college may require you) to take a co-requisite student success course called *Frameworks for Mathematics and Collegiate Learning.*

Structure of the curriculum

The curriculum is designed in 25-minute learning episodes, which can be pieced together to conform to any class length. These short bursts of active learning, combined with whole class discussion and summary, will help you with memory retention.[1]

---

[1] Sources: Buzan, T. (1989). *Master your memory* (Birmingham: Typersetters); Buzan, T. (1989). *Use your head* (London: BBC Books); Sousa, D. (2011). *How the brain learns, 4th ed.* (Thousand Oaks, CA: Corwin); Gazzaniga, M., Ivry, R. B., & Mangun, G. R. (2002). *Cognitive neuroscience: The biology of the mind, 2nd ed.* (New York: W.W. Norton); Stephane, M., Ince, N., Kuskowski, M., Leuthold, A., Tewfik, A., Nelson, K., McClannahan, K., Fletcher, C., & Tadipatri, V. (2010). Neural oscillations associated with the primary and recency effects of verbal working memory. *Neuroscience Letters, 473,* 172–177.); Thomas, E. (1972). The variation of memory with time for information appearing during a lecture. *Studies in Adult Education,* 57–62.

Prerequisite assumptions

The skills listed below will help you to succeed in *Foundations of Mathematical Reasoning*:

- Demonstrate procedural fluency with real number arithmetic operations (e.g., basic operations, comparing, contrasting), use arithmetic operations to represent real-world scenarios, and use those operations to solve stated problems.

- Use graphical representations on a real number line to demonstrate fluency when ordering numbers, to represent operations (e.g., addition, subtraction, doubling, halving, averaging), and to represent fractions and decimals.

- Demonstrate a basic understanding and familiarity with fractions, decimals, and percentages. Procedural competency for representing each number form and moving from one to the other is desired upon enrollment in this course, but you may also need to work with materials outside of class to review basic concepts and build basic skills.

Constructive perseverance level

You may or may not have experience with struggling in mathematics. Struggle is important, because struggling indicates learning. If struggle is not taking place, you are not being challenged and are not gaining new knowledge and skills. However, struggle that is unproductive often turns into frustration. The *Foundations of Mathematical Reasoning* course is designed to promote constructive perseverance. This means that you are supported in persisting through struggle.

The levels of constructive perseverance, outlined below, should be viewed as a broad continuum, rather than as distinct, well-defined categories. Some content requires greater structure and more direct instruction. The level is based both on the development of the students and the demands of the content. The levels are not designated in the student materials, but the descriptions may help you understand the structure of the lessons. The levels of constructive perseverance are as follows:

- **Level 1:** The problem is broken into sub-questions that help develop strategies. You and your fellow group members will reflect on and discuss questions briefly and then come back together to discuss with the whole class. This process moves back and forth between individual or small-group discussion and class discussion in short intervals.

  Role of the instructor: To help the class to develop the culture of discussion, establish norms of listening, and model the language used to discuss quantitative concepts.

- **Level 2:** The problem is broken into sub-questions that give you some direction but do not explicitly define or limit strategies and approaches. You will work in groups on multiple steps for longer periods, and the instructor will facilitate individual groups, as needed. The instructor brings the class together at strategic points at which important connections need to be made explicit, or when breakdowns of understanding are likely to occur.

Role of the instructor: To support you in working more independently and evaluating your own work so you feel confident about moving through multiple questions without constant reinforcement from the instructor.

- **Level 3:** The problem is not broken into steps or is broken into very few steps. You are expected to identify strategies for yourself. Groups work independently with facilitation by the instructor, as necessary. Groups report on results, and class discussion focuses on reflection of the problem as a whole.

    Role of the instructor: To support you in persisting with challenging problems, including trying multiple strategies before asking for help.

The role of the preview and practice assignments

One of the most important aspects of the *Foundations of Mathematical Reasoning* curriculum is the role and design of the homework assignments. These assignments differ from traditional homework in several ways:

- Each question has a specific purpose. While some questions are specifically skill based, repetition of a skill in a single form is never used. If repetition is deemed valuable, it is done with different contexts that require you to think about each question, rather than assume you can repeat the steps of the previous question.

- The preview assignments are designed to prepare you for the next lesson. This preparation is done explicitly. You are given a prerequisite set of skills, each of which is used in the next lesson. At the end of the assignment, you are asked to rate your ability to perform those skills. Be honest when rating yourself—if you are unprepared for class, you will be unable to participate fully. Forestall problems by seeking help before class on any skills you rate lower than a 4. The rating from the first preview assignment is shown below.

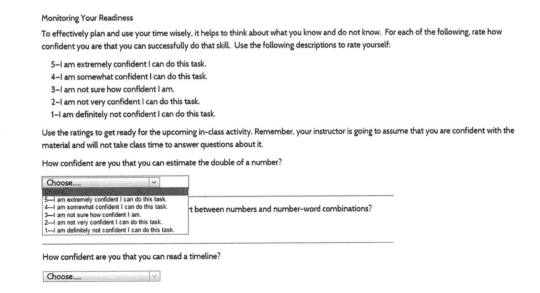

Monitoring Your Readiness

To effectively plan and use your time wisely, it helps to think about what you know and do not know. For each of the following, rate how confident you are that you can successfully do that skill. Use the following descriptions to rate yourself:

    5—I am extremely confident I can do this task.
    4—I am somewhat confident I can do this task.
    3—I am not sure how confident I am.
    2—I am not very confident I can do this task.
    1—I am definitely not confident I can do this task.

Use the ratings to get ready for the upcoming in-class activity. Remember, your instructor is going to assume that you are confident with the material and will not take class time to answer questions about it.

How confident are you that you can estimate the double of a number?

Choose.....   ⌄

5—I am extremely confident I can do this task.
4—I am somewhat confident I can do this task.
3—I am not sure how confident I am.   rt between numbers and number-word combinations?
2—I am not very confident I can do this task.
1—I am definitely not confident I can do this task.

How confident are you that you can read a timeline?

Choose.....   ⌄

- The preview assignments occasionally contain information or questions that are directly used in the next lesson.

- The practice assignments allow you to develop and practice skills from the current lesson. This may include similar problems in a new context or an extension of the learning within the same context.

- One goal of the course is for you to engage increasingly in productive struggle. Therefore, the assignments are based on the same principle of constructive perseverance as the rest of the curriculum. Ideally, each assignment should offer entry-level questions that you should be able to complete successfully along with more challenging questions. You should make a valid attempt on each question, even if you are not sure about the work.

- In some cases, the assignments include actual instructional materials. It is expected that you will read these materials, as they are usually not presented directly in class. This information distinguishes the material from traditional textbooks in which the text is often assigned by instructors, but often only used by students as a reference.

Student Resources

The Student Resources are designed to be a starting point for course reference materials. Some are directly lifted from an activity because they contain material that may be useful to you later in the course. Others were created as supplementary material.

Language and literacy skills

Quantitative literacy has unique language demands that are different from other subjects, even other math courses.

The learning outcomes of the course include the following:

- Reading and interpreting quantitative information from a variety of real-world sources.

- Communicating quantitative results both in writing and orally using appropriate language, symbolism, data, and graphs.

The designers of the course have further defined the expectations and purpose of reading and writing in the *Foundations of Mathematical Reasoning* course:

- You will read and use authentic text, which is defined as text that comes from a real-world source, or has been written by an author to replicate a real-life source. Using authentic text directly affects engagement and thus the development of skills in reading quantitative information in a variety of real-world situations.

- You will write in class and for assignments in order to:

- o   Make sense of quantitative information and processes, especially in relationship to a context.

- o   Develop skills in communicating quantitative information.

- o   Provide one form of assessment by which you may demonstrate your understanding of the course material.

Curriculum Design Standards

This course is the first step in a pathway to and through college mathematics. Here's how it works:

Standard I: Structure and Organization of Curricular Materials

This course is organized around big mathematical and statistical ideas and concepts as opposed to skills and topics.

Standard II: Active Learning

This course is designed to actively involve you in doing mathematics and statistics, analyzing data, constructing hypotheses, solving problems, reflecting on your work, and learning and making connections.

Class activities provide regular opportunities for you to actively engage in discussions and tasks using a variety of different instructional strategies (e.g., small groups, class discussions, interactive lectures).

Standard III: Constructive Perseverance

This course supports you in developing the tenacity, persistence, and perseverance necessary for learning mathematics.

Standard IV: Problem Solving

This course supports you in developing problem-solving skills, and in applying previously learned skills to solve nonroutine and unfamiliar problems.

Standard V: Context and Interdisciplinary Connections

This course presents mathematics and statistics in context and connects mathematics and statistics to various disciplines.

Standard VI: Use of Terminology

This course uses discipline-specific terminology, language constructs, and symbols to intentionally build mathematical and statistical understanding and to ensure that terminology is not an obstacle to understanding.

Standard VII: Reading and Writing

This course develops your ability to communicate about and with mathematics and statistics in contextual situations appropriate to the pathway.

Standard VIII: Technology

This course uses technology to facilitate active learning by enabling you to directly engage with and use mathematical concepts. Technology should support the learning objectives of the lesson. In some cases, the use of technology may be a learning objective in itself, as in learning to use a statistical package in a statistics course.

Note: A more detailed description of the design standards is available on the Dana Center website at **http://www.utdanacenter.org/higher-education/new-mathways-project/new-mathways-project-curricular-materials/foundations-of-mathematical-reasoning-course/**

Learning goals

The following five learning goals apply to this *Foundations of Mathematical Reasoning* course, with the complexity of problem-solving skills and use of strategies increasing as you advance through your pathway.

In this course, we define the ways that the learning goals are applied and the expectations for mastery. The bullets below each of the five learning goals specify the ways in which each learning goal is applied in the *Foundations of Mathematical Reasoning* course.

This course is designed so that you meet the goals across the courses in a given pathway. Within a course, the learning goals are addressed across the course's content-based learning outcomes.

Communication Goal: You will be able to interpret and communicate quantitative information and mathematical and statistical concepts using language appropriate to the context and intended audience.

In this *Foundations* course, you will…

- Use appropriate mathematical language.

- Read and interpret short, authentic texts such as advertisements, consumer information, government forms, and newspaper articles containing quantitative information, including graphical displays of quantitative information.

- Write 1 to 2 paragraphs using quantitative information to make or critique an argument or to summarize information from multiple sources.

Problem Solving Goal: You will be able to make sense of problems, develop strategies to find solutions, and persevere in solving them.

In this *Foundations* course, you will...

- Solve multi-step problems by applying strategies in new contexts or by extending strategies to related problems within a context.

Reasoning Goal: You will be able to reason, model, and make decisions with mathematical, statistical, and quantitative information.

In this *Foundations* course, you will...

- Make decisions in quantitatively based situations that offer a small number of defined options. The situations will not be limited to contexts in which there is a single correct answer based on the mathematics (e.g., which buying plan costs less over time), but will include situations in which the quantitative information must be considered along with other factors.

- Present short written or verbal justifications of decisions that include appropriate discussion of the mathematics involved.

Evaluation Goal: You will be able to critique and evaluate quantitative arguments that utilize mathematical, statistical, and quantitative information.

In this *Foundations* course, you will...

- Identify mathematical or statistical errors, inconsistencies, or missing information in arguments.

Technology Goal: You will be able to use appropriate technology in a given context.

In this *Foundations* course, you will...

- Use a spreadsheet to organize quantitative information and make repeated calculations using simple formulas.

- Use the internet to find quantitative information on a given subject. The topics should be limited to those that can be researched with a relatively simple search.

- Use internet-based tools appropriate for a given context (e.g., an online tool to calculate credit card interest).

Content learning outcomes

The content learning outcomes include both mathematical and contextual topics in keeping with the role of this course as a quantitative literacy course. The topics for the *Foundations of Mathematical Reasoning* course are:

- Numeracy

- Proportional Reasoning

- Algebraic Competence, Reasoning, and Modeling

- Probabilistic Reasoning to Assess Risk

- Quantitative Reasoning in Personal Finance

- Quantitative Reasoning in Civic Life

Numeracy

Outcome: You will develop number sense and the ability to apply concepts of numeracy to investigate and describe quantitative relationships and solve real-world problems in a variety of contexts.

You will be able to:

N.1   **Demonstrate operation sense and communicate verbally and symbolically with real numbers.**

For example: Know when and how to perform arithmetic operations with the use of technology. Use order of operations to identify an error in a spreadsheet formula. Predict the effects of multiplying a number by a number between 0 and 1.

N.2   **Demonstrate an understanding of fractions, decimals, and percentages by representing quantities in equivalent forms, comparing the size of numbers in different forms and interpreting the meaning of numbers in different forms.**

For example: Write a fraction in equivalent decimal form and vice versa. Compare growth expressed as a fraction versus as a percentage. Interpret the meaning of a fraction. Interpret the meaning of percentages greater than 100% and state if such a percentage is possible in a given context.

N.3   **Solve problems involving calculations with percentages and interpret the results.**

For example: Calculate a percentage rate. Explain the difference between a discount of 30% and two consecutive discounts of 15%. Calculate relative change and explain how it differs from absolute change.

N.4   **Demonstrate an understanding of large and small numbers by interpreting and communicating with different forms (including words, fractions, decimals, standard notation, and scientific notation) and compare magnitudes.**

For example: Compare large numbers in context, such as the population of the US compared to the population of the world. Calculate ratios with large numbers such as water use per capita for a large population. Interpret a growth rate less than 1%.

N.5   **Use estimation skills, and know why, how, and when to estimate results.**

For example: Identify and use numeric benchmarks for estimating calculations (e.g., using 25% as an estimate for 23%). Identify and use contextual benchmarks for comparison to other numbers (e.g., using US population as a benchmark to evaluate reasonableness of statistical claims or giving context to numbers). Check for reasonableness using both types of benchmarks.

N.6   **Solve problems involving measurement including the correct use of units.**

For example: Identify the appropriate units for perimeter, area, and volume. Calculate the amount of paint needed to paint a non-rectangular surface. Interpret measurements expressed in graphical form.

N.7   **Use dimensional analysis to convert between units of measurements and to solve problems involving multiple units of measurement.**

For example: Convert between currencies. Calculate the cost of gasoline to drive a given car a given distance. Calculate dosages of medicine.

N.8   **Read, interpret, and make decisions about data summarized numerically (e.g., measures of central tendency and spread), in tables, and in graphical displays (e.g., line graphs, bar graphs, scatterplots, and histograms).**

For example: Critique a graphical display by recognizing that the choice of scale can distort information. Explain the difference between bar graphs and histograms. Explain why the mean may not represent a typical salary.

Proportional Reasoning

Outcome: You will use proportional reasoning to solve problems that require ratios, rates, proportions, and scaling.

You will be able to:

PR.1 Represent, and use ratios in a variety of forms (including percentages) and contexts.

For example: Interpret a rate of change within a context using appropriate units. Interpret a percentage as a number out of 1,000. Compare risks expressed in ratios with unequal denominators (e.g., 1 in 8 people will have side effects versus 2 in 14).

PR.2 **Determine whether a proportional relationship exists based on how one value influences another.**

For example: Simple versus compound interest. Analyze whether an estimated percentage is reasonable based on proportions. Analyze the effects of scaling and shrinking that are proportional and non-proportional (e.g., the impact of changing various dimensions on perimeter, area, and volume).

PR.3 **Analyze, represent, and solve real-world problems involving proportional relationships, with attention to appropriate use of units.**

For example: Use individual water-use rates to predict the water used by a population. Use the Consumer Price Index to compare prices over time. Use a scale to calculate measurements in a graphic.

Algebraic Competence, Reasoning, and Modeling

Outcome: You will transition from specific and numeric reasoning to general and abstract reasoning using the language and structure of algebra to investigate, represent, and solve problems.

You will be able to:

A.1 **Demonstrate understanding of the meaning and uses of variables as unknowns, in equations, in simplifying expressions, and as quantities that vary, and use that understanding to represent quantitative situations symbolically.**

For example: Understand the different uses of variables and the difference between a variable and a constant. Be able to use variables in context and use variables as placeholders, as in formulas. Write an algebraic expression to represent a quantity in a problem. Combine simple expressions. Use notation with variables (e.g., exponents, subscripts) in simple and moderately complex expressions.

A.2 **Describe, identify, compare, and contrast the effect of multiplicative or additive change.**

For example: Compare and contrast the rate of change and/or behavior of a linear and an exponential relationship in context. Recognize that a multiplicative change is different from an additive change. Explain how the rate of change of a linear relationship differs from an exponential rate of change, as well as the ramifications of exponential change (growth can be very slow for a time but then increase rapidly).

A.3 **Analyze real-world problem situations, and use variables to construct and solve equations involving one or more unknown or variable quantities.**

For example: Demonstrate understanding of the meaning of a *solution*. Write a spreadsheet formula to calculate prices based on percentage mark-up. Solve a formula for a given value. Identify when there is insufficient information given to solve a problem.

A.4   Express and interpret relationships using inequality symbols.

For example: Use inequalities to express the relationship between the probabilities of two events or the size of two groups. Interpret a histogram based on intervals expressed with inequality symbols.

A.5   Construct and use mathematical models to solve problems from a variety of contexts and to make predictions/decisions.

Representations will include linear and exponential contexts.

For example: Given a statement of how the balance in a savings account grows because of monthly interest, construct a table of months and balances and then write a mathematical model that provides the balance for a given month.

A.6   Represent mathematical models in verbal, algebraic, graphical, and tabular form.

For example: Be able to move from any one representation to another. Given an initial value and information about change, create a table, graph, and/or algebraic model. Given an algebraic model, create a table of values.

A.7   Recognize when a linear model is appropriate and, if appropriate, use a linear model to represent the relationship between two quantitative variables.

For example: Given a set of data, make an informal, intuitive evaluation of the applicability of a linear or exponential model with a focus on recognizing the limitations of the model and identifying an appropriate domain for which the model might be used to make accurate predictions. Describe the rate of change using appropriate units: slope for linear relationships, or average rate of change over an interval for nonlinear relationships.

Probabilistic Reasoning to Assess Risk

Outcome: You will understand and critically evaluate statements that appear in the popular media (especially in presenting medical information) involving risk and arguments based on probability.

You will be able to:

R.1   Interpret statements about chance, risk, and probability that appear in everyday media (including terms like unlikely, rare, impossible).

For example: Interpret statements such as "for a certain population the risk of a particular disease is 0.005". Compare incidences of side effects in unequal group sizes.

R.2   Identify common pitfalls in reasoning about risk and probability.

For example: Identify inappropriate risk statements, such as when the size of reference groups is unknown (e.g., California, 2009, 88% of motorcycle accident fatalities were helmeted, 12% unhelmeted).

R.3   Interpret in context marginal, joint, and conditional relative frequencies in context for data summarized in a two-way table and identify which relative frequency is appropriate to answer a contextual question.

For example: Distinguish between reported relative frequencies that are marginal, joint, or conditional. Choose the relative frequency that is the most informative for a given purpose. Choose the appropriate direction of conditioning for a given context (the chance of cancer given a positive test result is not the same as the chance of a positive test result given cancer).

R.4   Demonstrate understanding of absolute risk and relative risk (percentage change in risk) by describing how each provides different information about risk.

For example: Interpret the different information conveyed when comparing the magnitude of the absolute risks and percentage change in risk (e.g., an 80% increase in risk associated with taking a particular medication could mean a change in risk from 0.001 to 0.0018 or from 0.1 to 0.18).

Quantitative Reasoning in Personal Finance

Outcome: You will understand, interpret, and make decisions based on financial information commonly presented to consumers.

You will be able to:

PF.1   Demonstrate understanding of common types of consumer debt and explain how different factors affect the amount that the consumer pays.

For example: Calculate the interest paid on credit card debt based on a credit score; explain how the length of the pay-off period affects the total interest paid; demonstrate the relationship between a percentage rate and the amount of interest paid; define basic terminology such as principal, interest rate, balance, minimum payment, etc.

PF.2   Demonstrate understanding of compound interest and how it relates to saving money.

For example: Demonstrate the different impacts of the saving period and the amount saved on the accumulated balance; use a given formula to calculate a balance; demonstrate an understanding of the meaning of a compounding period and use the appropriate terminology for different periods (e.g., quarterly, annually, etc.).

PF.3 Identify erroneous or misleading information in advertising or consumer information.

For example: Explain why statements about "average" benefits of a product such as a weight loss plan are misleading; identify misleading graphs that create an appearance of greater impact than is warranted.

Quantitative Reasoning in Civic Life

Outcome: You will understand that quantitative information presented in the media and by other entities can sometimes be useful and sometimes be misleading.

You will be able to:

CL.1 Use quantitative information to explore the impact of policies or behaviors on a population. This might include issues with social, economic, or environmental impacts.

For example: Calculate the effects of a small decrease in individual water use on the amount of water needed by a large population over time; determine if the minimum wage has kept pace with inflation over time.

CL.2 Identify erroneous, misleading, or conflicting information presented by individuals or groups regarding social, economic, or environmental issues.

For example: Explain how two statements can be both contradictory and true (e.g., the "average" amount of a tax cut expressed in terms of the mean and the median); identify how two pie charts representing different populations can be misleading.

A million, a billion, a trillion, a quadrillion—it is easy to lose a sense of the size of large numbers. So just how big is a billion?

1) Jot down anything you know about the number one billion. Then share with at least two neighbors.

2) How many YouTube videos do you think have over one billion views?

3) If a billion people stood shoulder-to-shoulder, how long would the line be? Make a prediction.

Credit: Andreser/Shutterstock

Objectives for the lesson

You will understand that:

☐ Large numbers can be represented in various ways.

☐ Collaborating with others can enhance learning.

You will be able to:

☐ Scale measurements of groups to represent individual elements

☐ Scale measurements to represent larger quantities of individual elements.

Stand in line shoulder-to-shoulder with your classmates.

4) How many people are in the line?

5) To the nearest inch, how long is the line?

6) What is the average shoulder width of the people in your line? How do you know?

7) How long would the line be if there were 1 thousand (1,000) people?

8) How long would the line be if there were 1 million (1,000,000) people?

9) How long would the line be if there were 1 billion (1,000,000,000) people?

Syllabus quiz

Student Name _____

1) What is the course instructor's office phone number?

2) On what day(s) and time(s) are office hours held?

3) What is the attendance policy for the course?

4) What overall percentage will result in a C grade for the semester?

5) True or False? For any false statement, please correct the statement.

   a) Attendance is crucial for success in this course.

   b) You will complete the entire midterm and the entire final exam in class.

   c) Late assignments can be handed in up to three weeks after the due date.

   d) Any student who attends class, takes good notes, completes assignments, and studies outside of class should receive a good grade in this course.

6) What are the criteria for receiving maximum points for participation?

7) Homework assignments account for what portion of your final grade?

8) When are the tests scheduled for this class?

Lesson 1, Part C, How big is a billion? (continued) Theme: Civic Life

Recall that you measured the length of a line of your classmates using inches in Lesson 1, Part A. Inches are a very small unit of measurement, appropriate for gauging distances smaller than the span of your arms. Larger distances require larger units of measurement.

In the United States, we use a customary system, where the major distance units are in inches (smallest), feet, yards, and miles (largest). Switching between these distance units is called converting units or unit conversion. Can you recall any of the unit conversions below? Work with a partner to fill in the blanks.

1)   _____ inches = 1 foot

2)   _____ feet = 1 yard

3)   _____ yards = 1 mile

4)   _____ feet = 1 mile

Objectives for the lesson

You will understand that:

- ☐ Different unit systems can change the way numbers look.
- ☐ Thinking about whether an answer makes sense can help your understanding.

You will be able to:

- ☐ Convert between units of measure.
- ☐ Calculate quantities in billions.

Recall when you measured the line with your classmates.

5)   To the nearest inch, how long was the line?

6)   What type of mathematical operation would you use to convert inches into feet (add, subtract, multiply, or divide)?

7)   How long is the line to the nearest tenth of a foot?

8)   Think about your answer. Should the number be larger or smaller than when you measured in inches? If your answer makes sense, enter the answer in the table below.

9)   What is the average shoulder width of the people in your line measured in feet?

10)   How long would the line be if it contained 1,000 people? How do you know?

11)   How long would that line be if measured in miles? (Round to the nearest hundredth.)

12) Continue filling in the table to help you determine the length of a line of one billion people. Notice how the numbers look different depending on what units you use to make the estimation.

Number of people	Length of line (in feet)	Length of line (in miles)
Your group = _____		
1 person		
1,000 people		
1,000,000 people (1 million)		
1,000,000,000 people (1 billion)		

Lesson 2, Part A, Doubling population   Theme: Civic Life

The **doubling time** of a population is the amount of time it takes a population to double in size. Calculating doubling time helps you understand how fast a population is growing. The table below gives historical estimates of the human population.

Population Estimates Throughout History[1]

Year	World Population (Lower bound, in millions)	Year	World Population (Lower bound, in millions)
10,000 BCE	1	1800	978
9,000 BCE	3	1850	1,262
8,000 BCE	5	1900	1,650
7,000 BCE	7	1950	2,519
6,000 BCE	10	1960	2,982
5,000 BCE	15	1970	3,692
4,000 BCE	20	1980	4,435
3,000 BCE	25	1985	4,831
2,000 BCE	35	1990	5,263
1,000 BCE	50	1995	5,674
500 BCE	100	2000	6,070
AD 1	200	2005	6,454
1750	791	2010	6,864

1)   What does the entry "8,000 BCE" mean? What was the world's population then?

---

[1] Some of this data were retrieved from the U.S. Census Bureau, www.census.gov/ipc/www/worldhis.html and www.census.gov/population/international/data/worldpop/table_history.php.

Objectives for the lesson

You will understand that:

☐ Growth can be measured in terms of doubling time.

☐ Doubling times can be used to compare growth during different periods.

You will be able to:

☐ Calculate quantities in the billions.

☐ Use data to estimate doubling time.

☐ Compare and contrast growth via doubling times.

2) How long did it take for the population to reach double that amount?

3) How long did it take for the population to double again?

4) Find at least three other examples of doubling population in the table. (It doesn't have to be exactly double.) For example, Earth's population doubled between 500 BCE and AD1, a period of 500 years. Record some notes so that you can explain your thinking.

5) Discuss your results with your group. Was the doubling time always the same? What does this tell you about the rate of growth of the human population over time?

6) Which of the following statements best describes the change in doubling times before 1750 AD?

   a) The doubling times generally decreased over time.

   b) Before 1750 AD, estimated population doubling times decreased from 2,000 years to 500 years.

   c) The doubling times decreased from 2,000 to 1,000.

7) Write a statement that describes the change in doubling times after 1800 AD.

8) What are some factors you think may have led to this change in doubling times?

9) Imagine that you are explaining the relationship of million, billion, and trillion to someone else. You may use words, symbols, and pictures. Your response should follow the guidelines found in Resource **Writing Principles**

Lesson 2, Part B, Scientific notation   Theme: Civic Life

At the beginning of this course, you worked with your class to answer the question "How big is a billion?" In reality, there are things so large or so numerous that thinking in billions is not large enough. Similarly, there can be things so small or rare, that thinking of them as one-billionth is still too big.

Credit: Steve Mann/Fotolia

For example, have you ever thought about how many different ways a deck of cards can be shuffled?

Objectives for the lesson

You will understand that:

☐ Very large and very small numbers can be represented in different ways.

☐ The way you represent a number can impact your understanding of what it means.

You will be able to:

☐ Represent large and small numbers in scientific notation.

☐ Convert numbers from scientific notation to standard notation.

There are 80658175170943878571660636856403766975289505440883277824000000000000 different ways to deal a 52-card deck.[2] This is quite a large number! In fact, it is so large that it is almost impossible to get an idea of how large it is just by looking at it. One way we express very large (or very small) numbers is by using **scientific notation**. To use scientific notation, we essentially think about the number using multiplication.

1) Refer back to Lesson 1, Part A. Assume that the average shoulder width of the people in the line was 1.325 feet. How long would the line be if it contained 10 million people? Express your answer in feet.

2) Round the answer to the nearest million. How many zeros are in the number?

3) Rewrite the answer using multiplication.

4) Think of a different way to express the number and write it down.

---

[2] Improbable things happen. In *RationalWiki*. Retrieved May 3, 2014, from http://rationalwiki.org/wiki/Improbable_things_happen.

Scientific Notation

A number in scientific notation is written in the form:

$a \times 10^n$ where $1 \le a < 10$; and n is an integer.

1 is included and 10 is not included

---

5)   Is the answer to question 4 in scientific notation? Why or why not? If not, express the answer in scientific notation.

6)   Use what you have learned to express the number of ways a 52-card deck can be shuffled in scientific notation.

7)   How does expressing this number in scientific notation change the way you understand the number?

The chances of an event happening can be described using probability. For example, there is a 1 in 2 chance of having a baby girl, which is the same as a probability of ½ or 0.5.

8)   The probability of having identical twins is approximately 35 ten-thousandths or 0.0035.[3] Express the number in scientific notation.

Sometimes, it's useful to use **standard notation** instead of scientific notation. To express a number in standard notation, just write out the number.

9)   Convert the number 7.56×10^4 from scientific form to standard form.

---

[3] Babycenter. (n.d.) Your chances of having twins or more. Retrieved May 3, 2014, from http://www.babycenter.com/0_your-likelihood-of-having-twins-or-more_3575.bc.

Lesson 2, Part C, Ratios in water use   Theme: Civic Life

In previous lessons, you analyzed how quickly Earth's population is growing. Rapid population growth could impact **sustainability**—that is, Earth's ability to continue to support human life.

Credit: patpitchaya/Fotolia

Objectives for the lesson

You will understand that:

- ☐ Ratios are one way to compare numbers of varying magnitudes.
- ☐ Ratios can be used to make different types of comparisons.
- ☐ Units should be included in ratios.

You will be able to:

- ☐ Recognize when it is appropriate to express a ratio as a percentage.
- ☐ Calculate ratios of large numbers.
- ☐ Estimate ratios of large numbers.

1) In 2011, the population of the United States was approximately 311 million and the world population was about 7 billion. Write a ratio comparing the U.S. population to the world population as a fraction. Think about how you could simplify the answer to use smaller numbers.

2) Since this ratio is a comparison of part of the world population to the total world population, the answer can also be written as a percentage. Calculate the percentage and write a contextual sentence.

3) In 2011, the population of China was 1.341 billion. Compare China's 2011 population to the world population with a ratio. Write your answer as a percent and as a fraction. Consider how many decimals to give in your final answer.

4) Compare China's population to the population of the United States. Write a sentence that interprets this ratio in the given context.

In the next activity, you will analyze how these growing populations may be affecting sustainability, by studying water footprints.

According to the website www.waterfootprint.org, "People use lots of water for drinking, cooking, and washing, but even more for producing things such as food, paper, cotton clothes, etc. . . . The **water footprint** of an individual, community, or business is defined as the total volume of freshwater that is used to produce the goods and services consumed by the individual or community or produced by the business."

The table below gives the population and water footprints of China, India, and the United States from 1997–2001.[4]

Population and Water Footprints

Country	Population	Total Water Footprint (in cubic meters per year)
China	1,257,521,000	883,390,000,000
India	1,007,369,000	987,380,000,000
United States	280,343,000	696,010,000,000

5) Complete the contextual sentence:

   The United States had about _____ million people during the period 1997–2001.

6) Write a contextual sentence describing the water footprint of the United States using a number word combination as in question 5.

7) Write the numbers in the table in scientific notation under the numbers written in standard notation.

You will continue working with this data in the next lesson.

---

[4] Retrieved May 3, 2014, from www.waterfootprint.org.

Lesson 2, Part D, Analyzing water footprints   Theme: Civic Life

In the previous activity, you read about water footprints. The **water footprint** of an individual, community, or business is defined as "the total volume of freshwater that is used to produce the goods and services consumed by the individual or community or produced by the business."

The table below gives the population and water footprints of China, India, and the United States from 1997–2001.[5]

Population and Water Footprints

Country	Population	Total Water Footprint (in cubic meters per year)	
China	1,257,521,000	883,390,000,000	
India	1,007,369,000	987,380,000,000	
United States	280,343,000	696,010,000,000	

1) Write the numbers in the table in scientific notation under the numbers written in standard notation.

Objectives for the lesson

You will understand that:
  ☐ Large numbers can be represented in many forms.
  ☐ Water use can be measured and compared as a water footprint for a country as a whole or as a per capita measure.

You will be able to:
  ☐ Write and interpret numbers in scientific notation.
  ☐ Perform operations with scientific notation using a calculator.

---

[5] Retrieved May 3, 2014, from www.waterfootprint.org.

2) Notice that the countries listed in the table are ranked from highest to lowest population. Using the data in the Total Water Footprint column, rank the countries (from highest to lowest) according to their total water footprint.

3) How much water did the average person use in the United States? Add a fourth column to your table to hold your answer and be prepared to justify your response.

4) The answer to the previous question is called the **water footprint per capita**. Place this label in the fourth column (including units) and then compute the water footprint per capita for China and India.

5) Rank the countries from highest to lowest water footprint per person.

6) The average person in the United States used how many times more water than the average person in China? Write your answer in a sentence and justify your response.

7) During the same period of time, how many times larger was the population of China compared with the population of the United States? Write your answer in a sentence and justify your response.

8) Write a couple of sentences that relate the information from the previous two questions and what these facts might mean in terms of **sustainability**

Lesson 3, Part A,	Theme: Civic Life
Large numbers in the media	**ANSWERS**

You are traveling down the highway and see a billboard with this message:

> # Crisis: 36% of the unemployed have been out of work for 27 weeks or more.

1) You do not see the name of the organization that put up the billboard. What groups might have wanted to publish this statement? What are some social issues or political ideas that this statement might support?

2) You hear and see commercials, billboards, and pamphlets using quantitative information every day. People use these statements as evidence to convince you to do certain things. What are some examples of things advertisements try to convince you to do?

You often do not know whether these statements are true. You may not be able to locate the information, but you can start by asking if the statement is reasonable.

Objectives for the lesson

You will understand that:

☐ Quantitative reasoning is a powerful tool in making sense of the world.

☐ Media reports often generate more questions than answers.

You will be able to:

☐ Evaluate reasonableness.

☐ Rewrite quantitative statements to improve clarity.

3) This data was published in January 2014 by the U.S. Bureau of Labor and Statistics[6] and declared a crisis by news outlets.[7] Do you think this was a reasonable statement to make?

---

[6] U.S. Department of Labor, Bureau of Labor Statistics. (2014.) News release: The employment situation — May 2014." Retrieved May 3, 2014, from http://www.bls.gov/news.release/pdf/empsit.pdf.

[7] Condon, S. (20140, February 7). Jobs figures underscore long-term employment 'crisis,' leaders say. Retrieved May 3, 2014, from http://www.cbsnews.com/news/jobs-figures-underscore-long-term-unemployment-crisis-leaders-say.

4)   Without doing any further research, how can you determine whether the statement is reasonable?

5)   In the previous question, you thought about different ways to decide if the statement was reasonable. One approach is to compare the total number of people who have been unemployed for 27 weeks to the total number of people who have found jobs.

For example, assume that you live in a town where at any given moment, precisely 100 people are out of work on any given week. Assume that the 36 people who were unemployed in week 1 remained unemployed for 27 weeks. Take a little bit of time to look over the table below.

Year	Total Number of Unemployed	Number Finding Jobs	Number Remaining Unemployed
Week 1	100	64	36
Week 2	100	64	36
Week 3	100	64	36
Week 4	100	64	36

.
.
.

Week 25	100	64	36
Week 26	100	64	36
Week 27	100	64	36

Part A: How many people found jobs over 27 weeks?

Part B: Of all the people who were unemployed for some time over the 27-week period, what percentage did not find a job at all?

Part C: What percentage of unemployed people in Week 27 were out of a job the entire time?

6) After doing this analysis, does it change your initial reaction to the statistic at the beginning of the lesson? Why or why not?

7) Is there a way to rewrite the statement to make it more representative of the data? Explain your reasoning.

8) Based on your own personal beliefs, how would you rewrite the statement?

Lesson 3, Part B, Seeking help   Theme: Student Success

In your life outside of school, what do you do when you get stuck on something? This could be anything, from trying to remember which spices you need to make your world-famous enchiladas, having to change the oil on your car, or having to make a serious life decision.

1)   List five things you've needed help with in recent memory. They don't have to be big things, but they can be. Think about problems you've had with issues like getting to school or work, getting the kids' homework done, getting the laundry done or the house clean, getting the car fixed, or figuring out what's for dinner. Or think about bigger issues, like deciding to go to college, getting married, or finding a new apartment or buying a house.

2)   Who do you go to when you need help? If you are willing to ask others for help in your daily life, are you also willing to ask others for help in your academic life? Why or why not?

Objectives for the lesson

You will understand that:
   ☐ Successful students ask for help.
   ☐ There are resources available for you to use.
   ☐ As members of a learning community, all students help one another.

You will be able to:
   ☐ Identify when you need to ask for help.

Sometimes students think they can't get help in math class because they believe none of their friends know math well enough to help them. But have you ever had a problem where just talking about it with someone made all the difference? Sometimes that can work with math as well.

3)   Many times in this class now, we have discovered concepts just by talking about them with each other. What concepts did you struggle with in this class initially, that were made easier to understand through discussions with your classmates?

4)   Talking about math outside of class can also help. How often have you met with someone from this class to complete homework assignments, preview assignments, or participate in other study activities? How did that help you? If you haven't met with any of your classmates outside of class, how could you reach out and begin talking to your classmates about math when you're not in class?

5)   In addition to getting informal help from others, there are also more formal ways to get help that are provided by your college. What resources did you learn about in your *Frameworks* class that you think will be helpful in this course? Have you used any of these resources yet?

6)   Have you found any other resources that would be helpful to other people in the class?

7)   Seeking help is important. It is also important to realize that you can offer help. What are some ways you might be able to help your classmates in this class?

8)   How do you know when you need help?

Lesson 3, Part C, Estimating sale prices   Theme: Personal Finance

You have been shopping for a new cell phone and you've found one on sale!

Regularly
$87.99

Now
20% off!

1)   What is your estimate of the sale price? Try to make your estimation calculations mentally. (You may write down your work if you need to, but do not use a calculator.)

2)   Discuss your strategy with your group. The group should have at least two different strategies to share with the class. Then document all of the group's strategies in your notes.

Objectives for the lesson

You will understand that:

☐ Standard benchmarks can be used in estimation.

☐ There are many strategies for estimating.

☐ Percentages are an important quantitative concept.

You will be able to:

☐ Use a few standard benchmarks to estimate percentages (i.e., 1%, 10%, 25%, 33%, 50%, 66%, 75%).

☐ Estimate the percent of a number, including situations involving percentages less than one.

Knowing some estimation strategies and percentage benchmarks allows you to make quick calculations when it is inconvenient or unnecessary to calculate exact results.

There are many good estimation strategies. It is important that you develop strategies that make sense to you. A strategy is wrong only if it is mathematically incorrect. You can refer to the Resource: **Rounding and Estimation** for examples of estimation strategies.

3)   Use some of the estimation strategies discussed in your group to estimate the price of the phone if the sale sign says:

Part A: 25% off

Part B: 35% off

Part C: 70% off

Estimations help you make calculations quickly in daily situations. This next problem shows how estimates of percentages can be used to make comparisons among groups of different sizes.

4)   A law enforcement officer reviews the following data from two precincts. She makes a quick estimate to answer the following question: "If a violent incident occurs, in which precinct is it more likely to involve a weapon?" Make an estimate to answer this question and document your strategy in your notes.

Precinct	Number of Violent Incidents	Number of Violent Incidents Involving a Weapon
1	25	5
2	122	18

5)   You have a credit card that awards you a "cash back bonus," which means that every time you use your credit card to make a purchase, you earn back a percentage of the money you spend. Your card gives you a bonus of 0.5%. Estimate your award on $462 in purchases. Document your strategy in your notes.

Lesson 3, Part D, Calculating sale prices   Theme: Personal Finance

Being able to calculate with percentages is also very important. In the last activity, an estimate of the crime data may help the officer decide about staffing, but if she is writing a formal report, she will need exact calculations.

Regularly
$87.99

Now
20% off!

1)   When might someone be satisfied with an estimate of the cell phone sale price and when might an exact calculation be needed?

Objectives for the lesson

You will understand that:

☐ Both estimation and exact calculation are valuable skills.

☐ There are multiple strategies for calculating percentages.

You will be able to:

☐ Determine when an estimate or an exact calculation is appropriate.

☐ Calculate the percent one number is of another.

☐ Calculate the percent of a number, including situations involving percentages less than one.

2)   Find the exact sale price if the phone is 20% off. Justify your answer.

3)   How close is your calculation to the estimate you made before?

Calculate the <u>exact</u> answers for the situations given. You may use a calculator; however, you still need to document your thinking on your paper.

4)   Find the sale price if the phone is:

Part A: 25% off

Part B: 35% off

Part C: 70% off

5)   Refer back to the table of crime reports from the previous activity. For each precinct, what is the exact percentage of incidents that involved a weapon? Round your calculation to the nearest whole percent.

6)   Calculate the exact amount of your "cash back bonus" if your credit card awards a 0.5% bonus and you charge $462 on your credit card.

7)   A $119.99 phone is on sale for $86. Estimate and then calculate the percentage discount, rounding to the nearest 1%.

Lesson 3, Part E, Developing self-regulation   Theme: Student Success

You have been doing self-assessment in the preview sections of your homework. How can you use this information to improve your learning?

Credit: wavebreakmedia/Shutterstock

Objectives for the lesson

You will understand that:

☐ Self-assessment is a skill that improves with practice and reflection.

☐ Self-regulation can help students study and learn more efficiently.

You will be able to:

☐ Evaluate the accuracy of your self-assessment to this point.

☐ Make a plan to continue to improve your self-assessment and use it to regulate your learning.

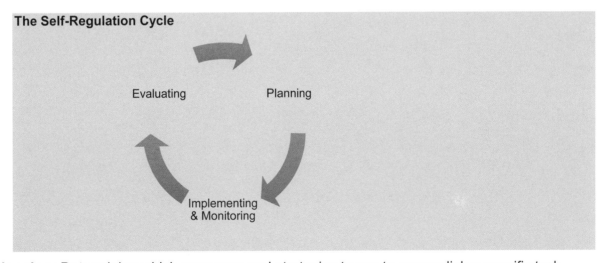

The Self-Regulation Cycle

Evaluating   Planning

Implementing & Monitoring

Planning: Determining which resources and strategies to use to accomplish a specific task.

Implementing & Monitoring: Executing your plan and continuously examining the progress you are making toward completing that specific task.

Evaluating: Assessing how well the planning and monitoring helped you complete the task.

1) How well have you been using your self-assessments in the assignments to prepare for class? Consider the following in answering this question. Give specific examples in your answer.

   Did your self-assessments match your performance? For example, if you rate yourself very low on every concept but you perform well, you are underestimating your understanding.

   Have you been thoughtful about your self-assessments? Sometimes people just check the same rating for every concept without really thinking about it.

   Have you used your self-assessments to review material in order to be prepared for class?

2) List strategies that you are currently using or will use in the future to prepare for class.

Lesson 4, Part A, Budgeting operations   Theme: Personal Finance

The table shown represents the budget of an average college student attending McLennan Community College.[8]

1) What does "room and board" mean? What do you think makes the difference between the room and board for living at home with parents versus living in an off-campus apartment?

Student Budget

	At home with parents	Off-campus apartment
Tuition and Fees	3,450	3,450
Books and Supplies	1,222	1,222
Room and Board	2,424	6,806
Transportation	2,935	2,413
Personal and Misc	2,088	1,819
Total	**12,119**	**15,710**

2) Approximately what percentage of the total cost of attendance does room and board cost when living at home with parents? Try to work without a calculator and show your steps.

Objectives for the lesson

You will understand that:

☐ Flexibility with calculations is an important quantitative skill.

☐ Different methods of calculation are often possible and helpful.

You will be able to:

☐ Write a calculation at least two different ways based on:

   o Equivalent forms of fractions/decimals.

   o Relationship of multiplication and division.

   o Properties and order of operations.

3) Write some notes about what you did to answer question 2. Be specific. If you worked in your head, what sequence of steps did you use?

4) Try to think of other ways to work the problem. Write them down, and be specific.

5) Discuss your strategies with your neighbor, making note of different techniques.

---

[8] McLennan Community College. (2013). Average cost of attendance – 2013–2014 academic year. Retrieved July 10, 2013, from http://www.mclennan.edu/financial-aid/docs/Average_Costs.pdf.

6)   Approximately what percentage of the total cost of attendance do books and supplies cost when:

Part A: living at home with parents?

Part B: living in an off-campus apartment?

7)   Does the difference in percentages mean that books and supplies cost more based on where a student chooses to live?

Below is an alternative way of presenting some of the data.

At home with parents

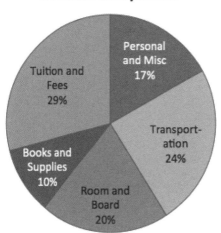

8)   What fraction could be used to approximate the portion of the budget spent on room and board?

9)   If a student has $1,160 to spend each month, how much is allocated to room and board? Try to work without a calculator and show your work.

10)   Discuss your strategy with your neighbor, making note of different techniques.

11)   Why is it important for the college to show students the actual values of the cost of attendance instead of percentages? How could potential students misinterpret percentages?

Lesson 4, Part B, Budgeting with spreadsheets   Theme: Personal Finance

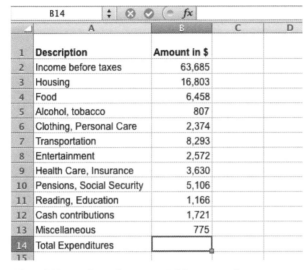

The table shown represents the financial information for the average family in 2011, according to the Consumer Expenditure Survey.[9]

1)   What does **expenditure** mean?

2)   What is an average family?

3)   What fraction could be used to approximate the portion of the family's income spent on food?

Objectives for the lesson

You will understand that:

☐ Technology is useful for managing and analyzing information.

You will be able to:

☐ Apply mathematical knowledge about properties and operations in using spreadsheet technology.

4)   A spreadsheet is a computer program used to organize and analyze data. In the example above, the <u>description</u> "Housing" is in cell A3. What cell contains the <u>amount</u> spent on "Entertainment"?

5)   Formulas can be used to perform calculations in spreadsheets, using the cell name as a variable. For example, what do you think the formula = B4 + B5 would represent for the data shown above?

---

[9] U.S. Bureau of Labor Statistics. (2012). Average annual expenditures and characteristics of all consumer units, Consumer Expenditure Survey, 2006–2011. Retrieved May 5, 2014, from http://www.bls.gov/cex/2011/standard/multiyr.pdf.

6)   Notice that the formula begins with an equal sign (=). This symbol is a requirement when entering a formula. What would the formula look like that would calculate a family's total expenditures? What cell should the result go in?

7)   What formula would calculate the difference between a family's income and its expenditures? Try to think of at least two different formulas that would work.

8)   What do you think happens to this "extra money" from question 7?

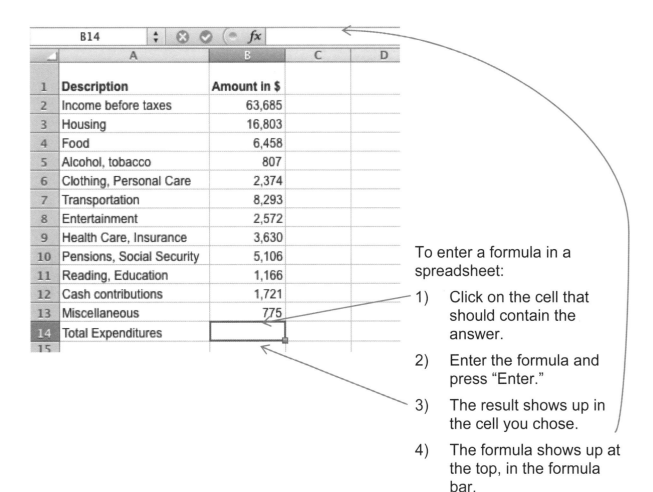

	A	B	C	D
	B14			
1	Description	Amount in $		
2	Income before taxes	63,685		
3	Housing	16,803		
4	Food	6,458		
5	Alcohol, tobacco	807		
6	Clothing, Personal Care	2,374		
7	Transportation	8,293		
8	Entertainment	2,572		
9	Health Care, Insurance	3,630		
10	Pensions, Social Security	5,106		
11	Reading, Education	1,166		
12	Cash contributions	1,721		
13	Miscellaneous	775		
14	Total Expenditures			
15				

To enter a formula in a spreadsheet:

1)   Click on the cell that should contain the answer.

2)   Enter the formula and press "Enter."

3)   The result shows up in the cell you chose.

4)   The formula shows up at the top, in the formula bar.

Lesson 4, Part C, Graph analysis   Theme: Personal Finance

1) With just a quick glance, what does this graph tell you about average household income?

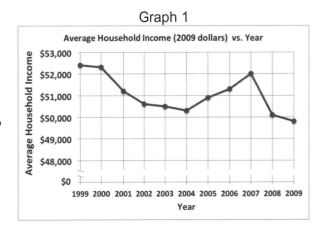

Graph 1

2) A period of growth in income occurred between which 2 years? Does it seem like a lot of growth?

Objectives for the lesson

You will understand that:

☐ The construction of a graph can change your perception of the information it represents.

You will be able to:

☐ Calculate absolute and relative change from a line graph.

Graph 2

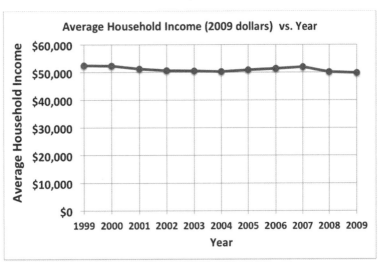

3) Now let's look at Graph 2. With just a quick glance, what does this graph tell you about average household income?

4) Does Graph 2 contradict the information shown in Graph 1? Explain.

5) What is the key difference in the two graphs? How does this affect the visual impression made by each graph?

6) Who might want to publish Graph 1 in the newspaper? Who might want to publish Graph 2?

You saw above how the visual impression made by the graphs was deceptive. This is why it is good to think about the actual data presented in graphs. One way to decide if there was "a lot" of change is to calculate the **relative change**.

In Preview Assignment 4.CD, **relative change** was defined as the ratio of "the amount of change" to "the original amount." In that assignment, you calculated the relative change over the entire period represented in the graph.

$$\text{Relative change in average household income} = \frac{\text{Ending amount} - \text{Beginning amount}}{\text{Beginning amount}}$$

(Note: Relative change is also sometimes stated as "amount of change **relative to** the original amount.")

7) What was the relative change in income from 2004 to 2007? Express the relative change rounded to the nearest one-tenth of a percentage point. Be sure your answer is labeled thoroughly in your notes.

8) The graph also shows a period of rapid decline in income. Determine the relative change in income for that period.

9) Write a contextual sentence about your answer to question 8. Refer to Resource **Writing Principles**, if needed.

Lesson 4, Part D, Using graphs to understand change

Theme: Civic Life

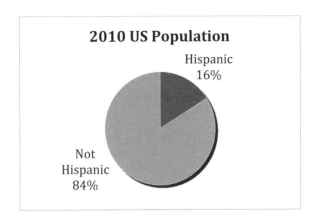

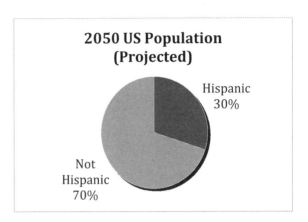

1) True or False: This pair of graphs predicts that the number of non-Hispanics in the United States is expected to decline between 2010 and 2050.

2) The U.S. population in 2010 was around 310 million. In 2050, the population is expected to be around 439 million.

   Part A: Estimate the number of Hispanic and non-Hispanic Americans for each of these years.

   Part B: Does your work support or refute the prediction you made in question 1?

Objectives for the lesson

You will understand that:

☐ To fully understand some graphs, the base value must be known.

You will be able to:

☐ Estimate the absolute size of the portions of a pie chart if given the base value.

☐ Use data displayed on two graphs to estimate a third quantity.

3) One newscast reported that the Hispanic population is expected to grow by 14% by 2050. Another station reported that it is expected to grow by 165%. Which report is correct, or are both or neither correct? Explain.

Question 3 illustrates an interesting twist to absolute and relative change. If your original quantities are given as percentages, then the absolute change should be expressed as a change in **percentage points**, not a percent because a percent is a relative change. This is a common error often seen in the media.

Let's apply what we know to the U.S. Gross Domestic Product. **Gross Domestic Product** is the value of all goods and services produced in the country. The **U.S. National Debt** is the money the federal government owes (this is different from the **deficit**). Make sure you understand the information shown in each graph. The graphs present the same information, though in significantly different ways.

The first graph shows the U.S. national debt by year.

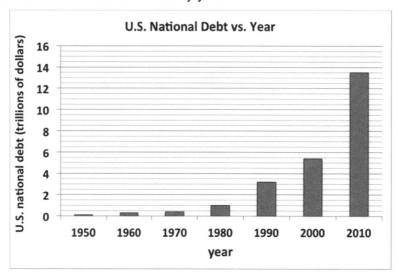

4)   Consider the statement, "The 2010 national debt is way out of hand and has never been higher." Do you agree or disagree? Explain.

The graph below shows the same information, but represents it as a percentage of the GDP.[10]

Federal Debt as a Percentage of GDP, 1790 to 2035 (projected)

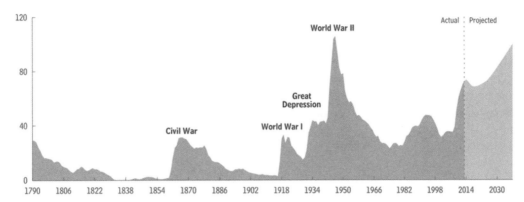

5) Consider the statement again: "The 2010 national debt is way out of hand and has never been higher." Do you still agree or disagree? Explain.

---

[10] Congressional Budget Office. (2013). The 2013 long-term budget outlook. Retrieved July 10, 2013, from http://cbo.gov/sites/default/files/cbofiles/attachments/44521-LTBO2013_0.pdf.

Lesson 5, Part A, Displaying table data   Theme: Personal Finance

Look back at the graphs you used in Lesson 4, Part D. The graphs showing Hispanic and non-Hispanic U.S. population illustrate one problem with several types of graphs: Once the graph has been made, you have lost the original data values.

A **stem-and-leaf plot** is one display that retains the original (or rounded) data values. Stem-and-leaf plots (also called stemplots) are based on place values. Take a look at some information on fuel economy collected by the U.S. Department of Energy.[11] (**Fuel economy** is another term for **gas mileage**, a measurement of how many miles can be driven per gallon of gas used.)

2013 Chevrolet Passenger Cars, Pickups, SUVs
Highway Miles per Gallon (Gas-Powered)

Stem	Leaf		Key: 1 5 means 15 mpg

```
Stem │ Leaf                              Key:   1│ 5 means 15 mpg
   1 │ 5 6 8 8 8 8 9 9
   2 │ 0 1 1 1 1 1 1 1 1 1 2 2 3 3 3 3 3 3 3 4 4 4 4 4 4 5 6 7 8 8 9 9 9
   3 │ 0 0 0 2 2 3 4 4 5 5 5 5 5 6 7 7 7 8 8 8 9
   4 │ 0 0 2
```

1)   Find the key for the stemplot shown.

   a)   The stems represent which place value?

   b)   Each leaf represents which place value?

Objectives for the lesson

You will understand:

   ☐ The format of a stem-and-leaf plot.

You will be able to:

   ☐ Read a stem-and-leaf plot and a back-to-back plot.

   ☐ Identify important features of a stem-and-leaf plot.

2)   Compare the number of "21" (numbers with 2 in the tens place) entries to the number of entries in the teens (numbers with 1 in the tens place). Use Resource **Writing Principles** to state your comparison in a complete sentence.

3)   How many 2013 gas-powered Chevy vehicles get more than 25 miles per gallon?

---

[11] U.S. Department of Energy. (n.d.) Retrieved May 15, 2014, from www.fueleconomy.gov/feg/bymodel/2013MakeList.shtml.

4)   Take turns in your group identifying interesting features in the plot. For example, "There are no 2013 gas-powered Chevrolets that get less than 15 mpg." Look for big ideas rather than individual entries. Record at least two observations from your group.

Stem-and-leaf plots can also be used to compare data sets. Here is an example with test grades for two American history classes.

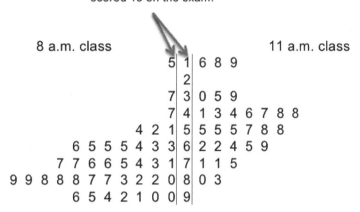

Someone in the 8 a.m. class
scored 15 on the exam.

8 a.m. class		11 a.m. class
	5	1 6 8 9
	2	
	7	3 0 5 9
	7	4 1 3 4 6 7 8 8
4 2	1 5	5 5 5 7 8 8
6 5 5 5 4 3 3	6	2 2 4 5 9
7 7 6 6 5 4 3 1	7	1 1 5
9 9 8 8 8 7 7 3 2 2 0	8	0 3
6 5 4 2 1 0 0	9	

5)   How many students scored in the 90s in the 8 a.m. class? In the 11 a.m. class?

6)   Once again, take turns in your group identifying interesting features in the plot. Look for big ideas rather than individual entries. Record at least two observations from your group.

7)   List the errors you see in the stem-and-leaf plot shown below.

9 a.m. English scores		10 a.m. English scores
	7 3	0 9 5
1 2 4	5	7 5 3 4 9 0
0 0 2 3 5 5 9 9	7	3 1 5
1 2 2	5 8	2 1
0 3 4 6 6 6	7 9	

Lesson 5, Part B,   Theme: Personal Finance
Relative frequency tables

Hwy MPG for 2013
Gas-Powered Cars
(Minicompacts)[12]

19	22	26	26
19	22	27	25
34	24	36	24
19	22	37	25
34	26	35	24
34	26	35	24
34	28	34	24
35	28	35	34
27	27	34	30
28	27	35	37
27	27	26	34
28	27	26	30
27	28	26	

The U.S. government tracks gas mileage by the size of the car. The table shown gives the highway gas mileage for 2013 cars in the minicompact size class.

1)   What information can you gain from a very quick glance at the table?

2)   If we wanted to make a stem-and-leaf plot of these data, what is the first thing we would have to do?

Another type of table, known as a **frequency distribution** or **frequency table**, helps to organize the data into a more useful form.

3)   What does the word *frequency* mean to you?

Objectives for the lesson

You will understand:

☐ The difference between **frequency** and **relative frequency**.

You will be able to:

☐ Choose appropriate bin size and sort data into bins.

☐ Report frequencies and relative frequencies.

---

[12] U.S. Department of Energy. (n.d.) Fuel economy of 2013 minicompact cars. Retrieved May 15, 2014, from www.fueleconomy.gov/feg/byclass/Minicompact_Cars2013.shtml.

4) To begin creating a frequency table, tally the data from the MPG table. The first three tallies from the data are illustrated for you.

Range	Tally	Frequency	
18 to 20	II		
21 to 23			
24 to 26			
27 to 29			
30 to 32			
33 to 35	I		
36 to 38			

5) The range column represents the **bins** that hold the data values; in this case, the bins are in increments of 3. The **frequency** represents the number of data values in each bin.

Part A: Label the third column "Frequency." What number is the first entry in the frequency column?

Part B: Write a sentence that explains what this number represents in context.

Part C: Complete the frequency column.

Part D: Check that the total of your frequency column equals the total number of data values.

6) Return to question 1. Does the frequency column help you form a better understanding of the gas mileage of 2013 minicompact cars? Explain.

7) **Relative frequency** is the percentage of all the data values in a particular bin.

Part A: The answer to question 5, part A represents what percent of all the data values? Round to the nearest tenth of a percent and justify your answer.

Part B: Write your answer to Part A (above) in sentence form.

Part C: Label the fourth column of the table "Relative Frequency." This column indicates how often each frequency occurs, when compared to the total. Share the work among your group to complete the rest of the relative frequency column.

Part D: What should be the total of the relative frequency column?

Part E: Explain the answer to question 7, part D with a complete sentence.

8)   Refer back to question 6 about the benefits to computing frequency. Does the <u>relative</u> frequency table help you form a better understanding of the gas mileage of 2013 minicompact cars? Explain.

9)   There is no specific rule for bin size or number of bins, but your choice of bin size can affect how the reader perceives the data. Let's explore what happens when you change the bin size.

Part A: Create a new relative frequency table with bins of size 10, as illustrated below.

Range	Tally	Frequency	Relative Frequency
15 to 24			
25 to 34			
35 to 44			

Part B: Does this new relative frequency table help you form a better understanding of the gas mileage data? Explain.

Lesson 5, Part C, Displaying data: Histograms

Theme: Personal Finance

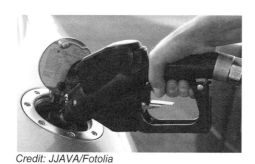

Credit: JJAVA/Fotolia

In an earlier activity, you created several frequency tables from the 2013 gas mileage data. While those tables helped to form a better understanding of the data, a picture can be even more helpful. A **histogram** is a graphical form of a frequency table.

Objectives for the lesson

You will understand that:

☐ Histograms display frequency and relative frequency.

You will be able to:

☐ Analyze histograms.

☐ Create a histogram from a frequency table.

1)   Refer to this frequency table with data from Lesson 5, Part B.

Ranges of Miles per Gallon (minicompacts)	Frequency
18 to 20	
21 to 23	
24 to 26	
27 to 29	
30 to 32	
33 to 35	
36 to 38	

Copy the following scale on your own paper and create a histogram by drawing bars with heights that represent the frequency values from the frequency table taken from Lesson 5, Part B. The first bar in the histogram is done for you. Make sure to label your histogram appropriately.

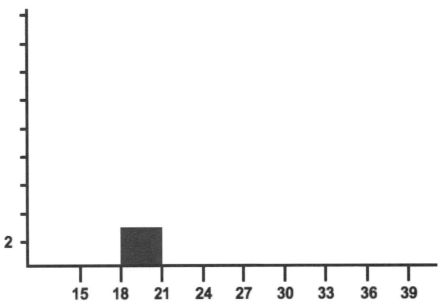

2)   Write a sentence explaining what the histogram illustrates. Does it help you gain a better understanding of the data than the frequency table did? If so, how?

3)   Now create another histogram, using the **relative frequencies** from the frequency table in question 1. The height of each bar represents the relative frequency for that bin (a percentage of the total). For instance, the three vehicles that get 18 to 20 miles per gallon represent almost 6% of the total data.

4)   Create a third histogram, using the frequency table below for the same data but with bins of size 10 instead of size 3.

Range	Frequency
15 to 24	
25 to 34	
35 to 44	

5)   Write a few sentences that compare and contrast the three histograms. Include comments on the shape and the information that you glean from them.

Lesson 5, Part D,
Shapes of distributions

Theme: Personal Finance

The graph shown below is called a **dotplot**. Each symbol on the plot represents a count for that measurement. For example, there is 1 person with a heart rate of 60 beats per minute, and 2 people with a rate of 100.

1) How would you describe the shape of this graph?

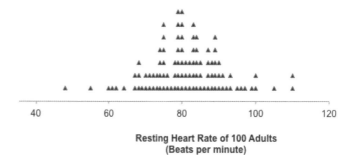

2) What does the graph tell you about adult heart rates?

Resting Heart Rate of 100 Adults
(Beats per minute)

Objectives for the lesson

You will understand that:

☐ Dotplots can be used to organize quantitative data for analysis.

☐ The shape of a distribution gives information about the data.

You will be able to:

☐ Analyze dotplots.

☐ Describe the shape of distributions.

3) Consider this new dotplot. Read it carefully to determine what the symbol means.

U.S. Census Bureau Report on 2011 Household Income

Household Income

✕ = approximately 1 million households. Data is rounded to nearest thousands place. Bin size = 5,000 so $50,000 means $50,000 to $54,000. (There is also a total of approximately 5 million households with $200,000 or more annual income.)[13]

---

[13] U.S. Census Bureau, Current population survey: 2011 household income. Retrieved May 16, 2014, from www.census.gov/hhes/www/cpstables/032012/hhinc/hinc01_000.htm.

Part A: Pick one column and write a sentence about its meaning. Refer to your Resource: **Writing Principles**, if needed.

Part B: What does the overall shape of this dotplot tell you about household incomes in the United States in 2011?

Part C: How would you describe the shape of this dotplot?

4)   Look at the stem-and-leaf from Lesson 5, Part A. How would you describe the distribution of this data?

**2013 Chevrolet Passenger Cars, Pickups, SUVs
Highway Miles per Gallon (Gas-Powered)**

Stem	Leaf		Key:	1 ∣ 5 means 15 mpg

Stem	Leaf
1	5 6 8 8 8 8 9 9
2	0 1 1 1 1 1 1 1 1 1 2 2 3 3 3 3 3 3 4 4 4 4 4 4 5 6 7 8 8 9 9 9
3	0 0 0 2 2 3 4 4 5 5 5 5 5 6 7 7 7 7 8 8 8 9
4	0 0 2

5)   How would you describe the shape of the frequency histogram from Lesson 5, Part C?

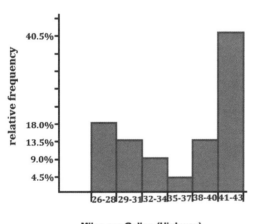

Fuel Economy of 2013 Diesel Vehicles

Miles per Gallon (Highway)

(Source: www.fueleconomy.gov/feg/byfuel/Diesel2013.shtml)

6)   Look at the relative frequency histogram for the fuel economy of 2013 diesel vehicles. How would you describe the shape of this histogram?

7)   What might explain the shape of the "Fuel Economy of 2013 Diesel Vehicles" histogram in question 5?

Lesson 6, Part A, Measures of central tendency

Theme: Personal Finance

Imagine that a sample of individuals report the balance on their credit card. The histogram summarizes this report.

1) How many individuals are included in the sample?

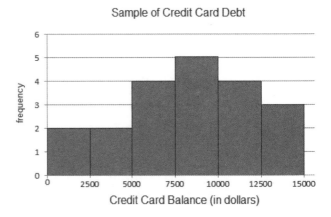

Sample of Credit Card Debt

Objectives for the lesson

You will understand that:

☐ Mean and median provide different pictures of data.

☐ Conclusions derived from statistical summaries are subject to misinterpretation.

☐ Histograms are graphical displays useful for showing the shape of a distribution.

You will be able to:

☐ Compare mean and median from the shape of a distribution.

☐ Create a data set that meets certain criteria for measures of central tendency.

---

2) Which bin contains the median credit card balance?

3) Create a data set that matches the histogram. In other words, record a set of dollar amounts including two values ranging from 0 to $2,499.99, two values ranging from $2,500 to $4,999.99, and so on.

4) Compute the mean and median of your data set.

5) Mark the mean and the median of your data set on the horizontal axis of the histogram.

6) Which phrase best describes the median of your data set?

    a) significantly less than the mean

    b) roughly the same as the mean

    c) significantly greater than the mean

Imagine another sample of individual credit card balances summarized by the histogram below.

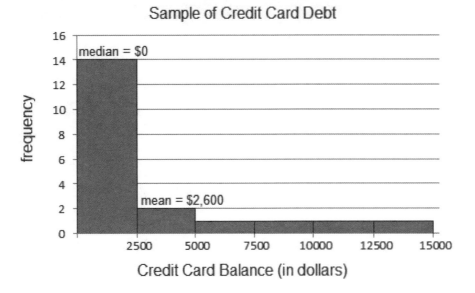

Sample of Credit Card Debt

7)   The median for this sample is $0. Interpret this statement. Explain what it means for the median to equal zero dollars.

8)   The mean for the sample is $2,600. Create a data set with a median of $0 and a mean of $2,600 that matches the histogram. In other words, write down a set of dollar amounts including fourteen values ranging from 0 to $2,499.99, two values ranging from $2,500 to $4,999.99, and one value in each of the remaining bins. Make sure that your 20 values have a median of $0 and a mean of $2,600.

9)   For convenience, name the four individuals in your data set who have credit card balances in the top four bins as the Spenders and the other individuals as the Savers. Here is a lame joke:

> The Savers are all at a party having a good time. When the Spenders arrive, the Savers suddenly get depressed. The Spenders say, "Hey, why is everyone suddenly so glum?" The Savers reply, "Don't you know what you four guys do to our average credit card debt?"

Explain the punchline of the joke. Why are the Savers depressed when the Spenders arrive?

Lesson 6, Part B, Brain power   Theme: Student Success

Many people believe that if you do not have a "math brain," you cannot learn math and if you do have a "math brain," learning math is easy.

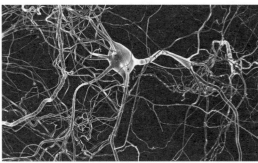

Credit: whitehoune/Fotolia

1) Do you think this is true? Why or why not?

Objectives for the lesson

You will understand that:

☐ Struggle, practice, and perseverance are key factors in learning.

You will be able to:

☐ Identify how the growth view of intelligence relates to learning math.

2) In your *Frameworks* course, you learned about the brain plasticity or how the brain grows. With your group, identify three things that you thought were the most important about this information.

3) How does this information relate to the idea of a "math brain"?

4) Which of the following two boxes contains the most dots?

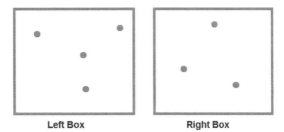

Left Box   Right Box

Did you have to count the dots or did your brain simply perceive that the left box contained more dots?

5)   Scientists are interested in knowing if substantially different neural processes govern perceiving differences in small quantities versus counting differences in larger quantities.[14] Which of the following two boxes contains the most dots?

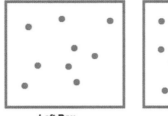 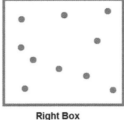

Left Box   Right Box

Did you have to count the dots this time?

6)   Do you sometimes feel that others "get" math the same way that you could tell the difference between four and three dots while you have to "work" at math in the same way that you had to count the difference between ten and nine dots? If so, you are not alone, but it may surprise you to learn that everyone has to work at learning math. What does brain plasticity suggest about learning math?

---

[14] Piazza, M., Mechelli, A., Butterworth, B., & Price, C.J. (2002). Are subitizing and counting implemented as separate or functionally overlapping processes? *Neuroimage, 15*(2), 435.

| Lesson 6, Part C, Making decisions with data | Theme: Personal Finance |

1) Examine the three advertisements shown below. Identify the measures of central tendency that appear in each ad. How are these measures of central tendency used in each ad?

Employment Opportunities

Sales Positions Available!	Are you above average?	NEED A NEW CHALLENGE?
We have immediate need for five enthusiastic self-starters who love the outdoors and who love people. Our salespeople make an average of $1,000 per week. Come join the winning team. Call 555-0100 now!	Our company is hiring one person this month—will you be that person? We pay the top percentage commission and supply leads. Half of our sales force makes over $3,000 per month. Join the *Above Average Team*! Call 555-0127 now! **We are!**	Join a super sales force and make as much as you want. Five of our nine salespeople each closed FOUR homes last month, and their average commission was $1,500 per sale. Do the math—this is the job for you. **Making dreams real—** **Call 555-0199**

Objectives for the lesson

You will understand that:

☐ Each statistic—mean, median, and mode—is a different measure of center for numerical data.

☐ You can use the measures of central tendency to make decisions.

You will be able to:

☐ Use data to make informed decisions.

☐ Interpret the mean, median, or mode in terms of the context of the problem.

☐ Match data sets with appropriate statistics.

2) For each advertisement, create a scenario that fits the information provided. This means to create a data set that fits the description, as you did in the previous activity with student credit card debt (Lesson 6, Part A).

3) In which job would you expect to earn the most money? Explain.

Lesson 6, Part D, Boxplots   Theme: Personal Finance

The data set shown is from a realtor report of the prices (in thousands) of all the homes sold in one month in a small city.

$550, $61, $75, $228, $79, $121, $79, $129, $240, $150, $147, $72, $142, $50

1)   Place the sales prices in order, from smallest to largest, in one long horizontal row on your paper.

2)   Do you notice anything interesting about this data set?

Objectives for the lesson

You will understand that:

☐ The 5-number summary can be used to describe numerical data.

☐ Conclusions derived from statistical summaries are subject to misinterpretation.

You will be able to:

☐ Analyze boxplots.

☐ Interpret the mean and median in terms of the context of the problem.

3)   Determine the **5-number summary**, which consists of the following values:

- o   the median

- o   the smallest number in the set (called the **minimum**)

- o   the largest number in the set (called the **maximum**)

- o   the median of the lower half (called Q_1)

- o   the median of the upper half (called Q_3)

4)   Create a number line like the one shown below. Mark each of the values from question 3 with a vertical hash mark. The Minimum of the data set is marked for you.

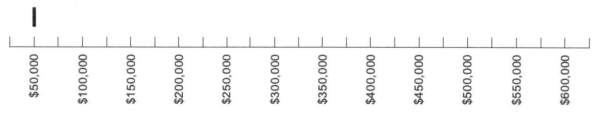

May Home Sales in Lakeside Village

5) How many of the data values are at or below the median value? What percent of the total number of values does this represent?

6) How many of the data values are at or below Q_1? What percent of the total number of values does this represent?

7) How many of the data values are at or above Q_3? What percent of the total number of values does this represent?

8) How many of the data values range from Q_1 to Q_3 (including Q_1 and Q_3)? What percent of the total number of values does this represent?

9) Find the mean of the data values to the nearest dollar.

10) Use the information from questions 5–9 to write a short news report about home sales.

Lesson 7, Part A, The credit crunch   Theme: Personal Finance

When you use a credit card, you can pay off the amount you charge each month. If you do not pay the full amount, you are borrowing money from the credit card company. The unpaid balance is credit card debt. Many people in the United States are concerned about the amount of credit card debt both for individuals and for society in general.

Credit: bernie_moto/Fotolia

In this lesson, you will use skills and ideas from previous lessons to think about some issues related to credit cards. You may want to refer back to the previous lessons.

Objectives for the lesson

You will understand that:

☐ Quantitative reasoning and math skills can be applied in various contexts.

☐ Creditworthiness affects interest rates and the amount paid by the consumer.

☐ Reading quantitative information requires filtering out unimportant information.

You will be able to:

☐ Apply skills and concepts from previous lessons in new contexts.

Annual Percentage Rate (APR) for Purchases	**0.00%** introductory APR for 6 months from the date of account opening. After that, your APR will be **10.99%** to **23.99%** based on your creditworthiness. This APR will vary with the market based on the Prime Rate.

Credit card companies measure **creditworthiness** with a credit score. High credit scores indicate good creditworthiness. Now you will explore how your credit score can affect how much you have to pay in order to borrow money. Juanita and Brian both have a credit card with the terms in the disclosure form given above and have had their cards for more than 6 months.

1)   Juanita has good credit and gets the lowest interest rate possible for this card. She is not able to pay off her balance each month, so she pays interest. Estimate how much interest Juanita would pay in a year if she maintained an average balance of $5,000 each month on her card. Explain your estimation strategy.

2)   Brian has a very low credit score and has to pay the highest interest rate. He is not able to pay off his balance each month, so he pays interest. Calculate how much interest he would pay in a year if he maintained an average balance of $5,000 each month. Show your calculation.

3)   What are some things that might affect your credit score?

Lesson 7, Part B, More credit crunch   Theme: Personal Finance

Refer to this credit card disclosure from Preview Assignment 7.A.

Credit:full image/Fotolia

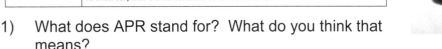

| Purchases | After that, your APR will be **10.99%** to **23.99%** based on your creditworthiness. This APR will vary with the market based on the Prime Rate. |

1) What does APR stand for? What do you think that means?

2) How often are you supposed to pay your credit card bill? How would you calculate the interest rate for that period of time?

Objectives for the lesson

You will understand that:

☐ Quantitative reasoning and math skills can be applied in various contexts.

☐ Creditworthiness affects interest rates and the amount paid by the consumer.

You will be able to:

☐ Apply skills and concepts from previous lessons in new contexts.

☐ Interpret a numerical expression in context.

☐ Write a spreadsheet formula to compute interest.

3) Juanita has good credit and gets the lowest interest rate possible for her credit card. She has a balance of $982 on her January statement. What is her periodic rate?

4) Which of the following is the best estimate of how much interest she will pay? Explain your answer.

Less than a dollar $5–$10 $10–$20   More than $20

A **cash advance** is when you use your credit card to get cash instead of using it to make a purchase. For most cards, cash advances are more expensive than purchases in two ways: 1) The interest rate is higher; and 2) Interest is charged on an advance from the first day it is made. Interest is charged only on purchases if you do not pay the full amount each month. Refer to this disclosure from Preview Assignment 7.A.

Purchases	After that, your APR will be **10.99%** to **23.99%** based on your creditworthiness. This APR will vary with the market based on the Prime Rate.

APR for Cash Advances	**28.99%**. This APR will vary with the market based on the Prime Rate.

5) Discuss each of the following statements. Decide if it is a reasonable statement. Be prepared to justify your responses.

Part A: Jeff pays the highest interest rate for purchases. He would pay $0.05 per dollar more in interest for a cash advance than for a purchase.

Part B: The interest for cash advances is a little more than two-and-a-half times as much as for the lowest rate for purchases.

6) Brian used a spreadsheet to record his credit card charges for a month.

◇	A	B	C
1	Credit Card Charges for the Month		
2	Gas	$56.08	
3	Groceries	$36.72	
4	Restaurant	$12.82	
5	Movie tickets	$16.00	
6			
7	Interest paid		
8			

Brian used the following expression to calculate his interest for these charges for one month.

$$= \frac{0.2399}{12} * (B2 + B3 + B4 + B5)$$

Which of the following statements provides the best explanation of what the expression means in terms of the context?

a) Brian added his individual charges. Then he divided 0.2399 by 12. Then he multiplied the two numbers.

b) Brian found the interest charge for the month by dividing 0.2399 by 12 and multiplying it by the sum of Column B.

c) Brian found the periodic rate by dividing the APR by 12 months and multiplied the periodic rate by the sum of the individual charges. This calculation gives the interest charge for the month.

7) What is another spreadsheet formula Brian could have used to calculate his interest charges?

Lesson 7, Part C, A taxing situation   Theme: Personal Finance

An IRS tax worksheet called the Short Schedule SE appears below. Self-employed individuals use the Short Schedule SE to calculate their self-employment tax. Scan the document to get a sense of the content and the process.

Section A—Short Schedule SE.   Caution. Read above to see if you can use Short Schedule SE.

1a Net farm profit or (loss) from Schedule F, line 36, and farm partnerships, Schedule K-1 (Form 1065), box 14, code A	**1a**	
b If you received social security retirement or disability benefits, enter the amount of Conservation Reserve Program payments included on Schedule F, line 6b, or listed on Schedule K-1 (Form 1065), box 20, code Y	**1b** (	)
2 Net profit or (loss) from Schedule C, line 31; Schedule C-EZ, line 3; Schedule K-1 (Form 1065), box 14, code A (other than farming); and Schedule K-1 (Form 1065-B), box 9, code J1. Ministers and members of religious orders, see page SE-1 for types of income to report on this line. See page SE-3 for other income to report	**2**	
3 Combine lines 1a, 1b, and 2. Subtract from that total the amount on Form 1040, line 29, or Form 1040NR, line 29, and enter the result (see page SE-3)	**3**	
4 Multiply line 3 by 92.35% (.9235). If less than $400, you do not owe self-employment tax; **do not** file this schedule unless you have an amount on line 1b . . . . ▶	**4**	
Note. If line 4 is less than $400 due to Conservation Reserve Program payments on line 1b, see page SE-3.		
5 **Self-employment tax.** If the amount on line 4 is: • $106,800 or less, multiply line 4 by 15.3% (.153). Enter the result here and on **Form 1040, line 56,** or **Form 1040NR, line 54** • More than $106,800, multiply line 4 by 2.9% (.029). Then, add $13,243.20 to the result. Enter the total here and on **Form 1040, line 56,** or **Form 1040NR, line 54**	**5**	
6 **Deduction for one-half of self-employment tax.** Multiply line 5 by 50% (.50). Enter the result here and on **Form 1040, line 27,** or **Form 1040NR, line 27** **6**		

For Paperwork Reduction Act Notice, see your tax return instructions.   Cat. No. 11358Z   Schedule SE (Form 1040) 2010

Objectives for the lesson

You will understand that:

☐ Order of operations is needed to communicate mathematical expressions to others.

You will be able to:

☐ Perform multi-step calculations using information from a real-world source.

☐ Rewrite multi-step calculations as a single expression.

1) Sundos Allianthi sells crafts, such as jewelry and baskets, for extra money. She does not have a farm or get any of the benefits on line 1b. In 2010, she sold $11,385 in crafts and her expenses totaled $3,862. Expenses are the things she bought for her business.

Fill out Section A—Short Schedule SE for Sundos. How much self-employment tax does Sundos owe? Assume that line 29 of her 1040 form has a 0 amount. This amount is asked for on line 3 of the Short Schedule SE. Round entries to the nearest penny (hundredths place).

2)   Companies match the FICA contributions of their employees, so the employer pays 7.65% and the employee pays 7.65%. Self-employed individuals do not have the advantage of an employer who matches their FICA contribution. To compensate self-employed individuals for this disadvantage, the law allows self-employed individuals to multiply their net profit by 92.35% (0.9235) to compute taxable net earnings (the amount on line 4). Why do you think the law uses 92.35%?

3)   What does the amount on line 5 represent?

4)   Write the calculations you used to find Sundos's tax as a single expression that someone else could use and understand.

Lesson 7, Part D, A taxing situation (continued)

Theme: Personal Finance

Review your work from Lesson 7, Part C, where you calculated Sundos's self-employment taxes. Then read the **Objectives for the lesson**, given below.

Objectives for the lesson

You will understand that:

☐ Order of operations is needed to communicate mathematical expressions to others.

You will be able to:

☐ Perform multi-step calculations using information from a real-world source.

☐ Generate an algorithm for performing multi-step calculations.

☐ Explain the meaning of a calculation within a context.

Complete question 1 individually. When you are finished, check your answer with your group.

Section A—Short Schedule SE.   Caution. Read above to see if you can use Short Schedule SE.

1a	Net farm profit or (loss) from Schedule F, line 36, and farm partnerships, Schedule K-1 (Form 1065), box 14, code A	**1a**	
b	If you received social security retirement or disability benefits, enter the amount of Conservation Reserve Program payments included on Schedule F, line 6b, or listed on Schedule K-1 (Form 1065), box 20, code Y	**1b** ( )	
2	Net profit or (loss) from Schedule C, line 31; Schedule C-EZ, line 3; Schedule K-1 (Form 1065), box 14, code A (other than farming); and Schedule K-1 (Form 1065-B), box 9, code J1. Ministers and members of religious orders, see page SE-1 for types of income to report on this line. See page SE-3 for other income to report	**2**	
3	Combine lines 1a, 1b, and 2. Subtract from that total the amount on Form 1040, line 29, or Form 1040NR, line 29, and enter the result (see page SE-3)	**3**	
4	Multiply line 3 by 92.35% (.9235). If less than $400, you do not owe self-employment tax; **do not** file this schedule unless you have an amount on line 1b . . . . ▶	**4**	
	Note. If line 4 is less than $400 due to Conservation Reserve Program payments on line 1b, see page SE-3.		
5	**Self-employment tax.** If the amount on line 4 is: • $106,800 or less, multiply line 4 by 15.3% (.153). Enter the result here and on **Form 1040, line 56, or Form 1040NR, line 54** • More than $106,800, multiply line 4 by 2.9% (.029). Then, add $13,243.20 to the result. Enter the total here and on **Form 1040, line 56, or Form 1040NR, line 54**	**5**	
6	**Deduction for one-half of self-employment tax.** Multiply line 5 by 50% (.50). Enter the result here and on **Form 1040, line 27, or Form 1040NR, line 27** **6**		

For Paperwork Reduction Act Notice, see your tax return instructions.   Cat. No. 11358Z   Schedule SE (Form 1040) 2010

1)   Raven Craig started a tutoring business at the end of 2010. She has no income to report on line 1a or line 1b of Schedule SE. She also has a 0 entry on line 29 of Form 1040. She earned $1,050 and her expenses totaled $630. How much self-employment tax does Raven Craig owe?

Questions 2–9: Look back at the expression you wrote in Lesson 7, Part C. Think of yourself as a tax expert working with a computer programmer. Your task is to explain how to compute an individual's self-employment tax when the individual knows his or her revenues and expenses. The individual has no farm profits (or losses) and no social security or disability benefits. The individual has a 0 entry for line 29 of Form 1040 (or Form 1040NR). You must word your explanations step-by-step to help the programmer understand the process fully.

2)   Assume the program has already established that the individual has 0 entries on lines 1a and 1b. The programmer wants the program to compute line 2. What information should the computer program request first?

3)   What information should the computer program request second?

4)   What operation should the program perform to compute the amount reported in line 2?

5)   Next, the programmer wants to compute line 3. Assume the program has already established a 0 entry for line 29 of Form 1040 (or Form 1040NR). What operation should the program perform to compute the amount reported in line 3? Remember to assume 0 entries in 1a and 1b.

6)   Next, the programmer wants to compute line 4. What operation should the program perform to compute the amount reported in line 4?

7)   Finally, the programmer wants to compute line 5. Help the programmer complete the if-then statements.

   If line 4 is less than 400, then . . .

   If line 4 is greater than 400 but less than $106,800, then . . .

   If line 4 is greater than $106,800, then . . .

8)   An algorithm is a set of rules or instructions for solving a problem. Summarize and combine your answers to problems 2 through 7 to write a step-by-step algorithm that others can follow to complete the Schedule SE up to line 5 (assuming no farm profits, no social security or disability benefits, and no entry on line 29 of Form 1040).

9)   Exchange your algorithm with that of another group. Use their algorithm to compute the self-employment tax for a self-employed individual who earned $1,596,250 in revenue with expenses totaling $76,250.

Lesson 8, Part A, What is the risk?   Theme: Risk Assessment

A well-known newspaper once reported that the risk of colorectal cancer is increased by 21% for every 1.7 ounces of daily processed meat consumption.

The media often presents the public with information like this about various health risks, but it can be difficult to interpret the meaning. In the next few lessons, you will explore information about risk.

Credit: krasyuk/Fotolia

Objectives for the lesson

You will understand that:

☐ Risk can be measured in absolute and relative terms.

☐ Contextual information is important in interpreting risk.

☐ Different ways of expressing data serve different purposes for communication.

You will be able to:

☐ Express and interpret ratios in non-standard forms (i.e., 1 out of 20) and standard forms, such as percentages and natural frequencies out of multiples of 10.

☐ Identify missing information needed to calculate a relative measure.

1) Recall that the lesson opening stated that the risk of colorectal cancer is increased by 21% for every 1.7 ounces of daily processed meat consumption. What risk measure does the information provide? What does this risk measure mean?

2) Consider the data for the United States reported by the American Cancer Society.[15]

Risk of Developing Colorectal Cancer in a Lifetime

Males	1 in 20
Females	1 in 22

What percent of American men develop colorectal cancer during their lifetime?

---

[15] American Cancer Society. (n.d.) Lifetime risk of developing or dying from cancer. Retrieved March 31, 2014, from http://www.cancer.org/cancer/cancerbasics/lifetime-probability-of-developing-or-dying-from-cancer.

3) The American Cancer Society reported the data about men as a fraction with a numerator of 1. In question 2, you reported the data as a percent. There are many other ways to report risk. For example, convert the percent to a fraction with a denominator of 100. This fraction is a **natural frequency** often reported as
"___ men out of 100 men will develop colorectal cancer in their lifetime."

4) The ratios reported by the American Cancer Society are baseline risk figures because they provide a basis for estimating the chance that an individual in the United States will develop colorectal cancer without knowing anything specific about the particular individual (other than their sex and that they live in the United States). What is the baseline risk for an American woman to develop colorectal cancer expressed as a percent?

5) Imagine that every American male begins eating an additional 1.7 ounces of bacon daily (bacon is a processed meat). According to the newspaper article, the risk for colorectal cancer increases by 21%. The current risk for developing cancer is 5%. Multiply 21% by 5%. What is the product expressed as a percent?

6) The product from question 5 represents the increase in baseline risk caused by eating an additional 1.7 ounces of bacon.

   Part A: Add this increased risk to the original baseline risk reported by the American Cancer Society and round to the nearest whole percent.

   Part B: Express this adjusted baseline risk as a natural frequency, "the number of population members out of 100" who will develop colorectal cancer in their lifetime.

7) Look at the information gathered in the previous question. How many additional men per 100 develop colorectal cancer in their lifetime due to eating more bacon daily?

8) The newspaper article reported a 21% increase in risk. This risk measurement is a relative risk since the 21% increase is relative to the existing 5% risk for men (and 4.5% for women). In question 5, you calculated the absolute increase in risk as a percent. In question 7, you calculated the absolute increase in risk as a natural frequency. In your opinion, which of these forms gives the "most useful" information—that is, which form communicates your personal risk most effectively?

9) Why do you think news reports report risk in relative terms?[16]

_____

[16] For further reading, see "Risk: Bring home the bacon" (pp. 97–110) in M. Blastland and A. W. Dilnot, *The numbers game*. New York: Penguin.

Lesson 8, Part B, An apple a day   Theme: Risk Assessment

In another article, the same newspaper linked eating white fruits (such as apples and pears) and white vegetables (such as cauliflower) with lowered risk of stroke.

Credit: schiros/Fotolia

The information came from a university study in the Netherlands that indicated that there is a 9% decrease in the risk of stroke for every 25 grams of daily white fruit and vegetable consumption.

As you saw in the previous lesson, risk figures often appear in the media, but news articles do not always present these risk measurements in meaningful ways. Indeed, the risk measures frequently appear in relative terms without providing baseline figures to help make sense of the risk. In this lesson, you will further explore how to make risk personal in terms that make sense to you.

Objectives for the lesson

You will understand that:
- ☐ There are important questions to ask in assessing data about risk.
- ☐ Risk can be measured in absolute and relative terms.

You will be able to:
- ☐ Identify major factors in determining risks of disease.
- ☐ Make quantitative comparisons between data.
- ☐ Write a statement that interprets data from a table using multiple categories.

1) Locate the data given about stroke risk in the lesson introduction. What risk measurement appears in the article? Does this risk measurement encourage or discourage the consumption of white fruits and vegetables?

2) As you saw in the previous lesson, relative risk figures make little sense without an estimate of the baseline risk. Here is the baseline risk for stroke: 1 in 6 individuals worldwide will have a stroke in their lifetime.[17] If this baseline figure is correct, what is the general lifetime risk of stroke for a person written as a percent (rounded to the nearest tenth) and as a natural frequency of "___ individuals out of 100 individuals"? Approximate with a whole number of individuals in the natural frequency.

3) Apples are white fruit, and a typical apple weighs about 75 grams. Recall from Preview Assignment 8.B that three consecutive 9% reductions lead to an overall approximate reduction of almost 25%. According to the newspaper article, what is the relative decrease in risk for stroke for a person who eats "an apple a day"?

4) Think about a general population of people who eat "an apple a day" but would otherwise have a general lifetime risk of stroke of 1 in 6. Given the information in the newspaper article, what is the decrease in lifetime risk of stroke for this population of apple eaters?

5) Think about your answers to questions 2 and 4. What is the approximate lifetime risk for stroke expressed as a natural frequency of "___ individuals out of 100 individuals" for the population of apple eaters?

6) About how many fewer people in every 100 people suffer a stroke during their lifetime in the population of apple eaters compared to the general worldwide population?

7) The table shows the lifetime risk of suffering a first-ever stroke (of any type) for individuals based on their blood pressure at age 65.[18] Hypertension is a clinical term for high blood pressure.

Lifetime Risk of Stroke for Individuals at Age 65

	Normal Blood Pressure	Hypertension
Men	10.4%	20.5%
Women	14.6%	25.8%

---

[17] American Heart Association. (2013). One in six people worldwide will have a stroke in their lifetime. Retrieved April, 2 2014, from http://newsroom.heart.org/news/one-in-six-people-worldwide-will-have-a-stroke-in-their-lifetime.

[18] Izzo, J. L., Sica, D. A., & Black, H. R. (2008). *Hypertension primer: Essentials of high blood pressure* (4th ed.) Dallas, TX: American Heart Association.

Convert the values in the previous table to natural frequencies out of 1,000 individuals. Write the natural frequencies in the table below. Part of the table is already completed for you.

Lifetime Risk of Stroke for Individuals at Age 65

	Normal Blood Pressure	Hypertension
Men	104 out of 1,000	
Women		

8) Complete the table of estimates for a population of individuals who regularly eat "an apple a day." Report each risk measurement as a natural frequency out of 1,000 individuals. Part of the table is already completed.

Lifetime Risk of Stroke for "An Apple a Day" Individuals at Age 65

	Normal Blood Pressure	**Hypertension**
Men	78 out of 1,000	
Women		

9) Look carefully at the categories in the tables in questions 7 and 8.

Part A: Write a sentence, making at least one comparison within one table.

Part B: Write a sentence, making at least one comparison between the two tables.

Lesson 8, Part C, Reducing the risk   Theme: Risk Assessment

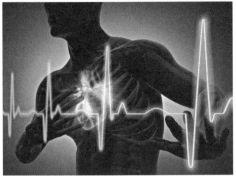

Many drugs are designed to reduce the risk of a disease or to reduce the severity. A new drug advertises that it "reduces the risk of heart attack by 50%."

1)   Do you think this is an effective or "good" drug?

Credit: psdesign1/Fotolia

Objectives for the lesson

You will understand that:

☐ The change in a quantity can be expressed as an absolute change and a relative change.

☐ There is often ambiguity in the English language when talking about the change of a quantity that is represented by a percent (e.g., many rates).

You will be able to:

☐ Create graphs that show absolute change calculated from a rate.

☐ Compute absolute changes.

In order to better understand the benefits of this drug, you will examine heart attack risk for two different groups.

Group 1 consists of individuals in Africa who:

- are 40 years old,
- do not have Diabetes Mellitus,
- smoke tobacco,
- have high cholesterol, and
- have high blood pressure.

Group 2 consists of individuals in Africa who:

- are 40 years old,
- do not have Diabetes Mellitus,
- do not smoke,
- have low cholesterol, and
- have high blood pressure.

The World Health Organization (WHO) reports that individuals in Group 1 have a greater than 40% chance (or risk) of suffering a heart attack within 10 years. The same report indicates that individuals in Group 2 have a less than 10% risk of suffering a heart attack within 10 years.[19]

---

[19] Source: World Health Organization, http://www.who.int/en.

2)   What are the differences between individuals in Group 1 and Group 2?

You will be assigned to work on either Group 1 (greater than 40% chance of heart attack) or Group 2 (less than 10% chance of heart attack). Wait for your assignment.

3)   If 500 people from your assigned group are observed for 10 years, how many individuals would you expect to suffer a heart attack within this time period?

4)   If 500 people from your assigned group are treated with the new drug from question 1 and then observed for 10 years, how many individuals would you expect to suffer a heart attack within this time period?

5)   Create a bar graph, using the one below as a guide, that represents the information from questions 3 and 4. Use different shading for the bars representing "with treatment" and "without treatment."

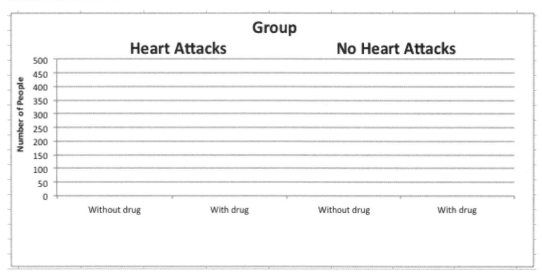

6)   You are planning to publish this report in a health blog to advise people interested in using the drug. Write a few sentences to accompany this graph.

Lesson 8, Part D, Is reducing the risk worth it?   Theme: Risk Assessment

Credit: michaeljung/Fotolia

Refer back to Lesson 8, Part C, in which a new drug reduces the risk of heart attack by 50%.

Suppose a group of people have a 0.5% chance of suffering a heart attack. How do you think the absolute and relative changes in the number of individuals (out of 500) who are expected to suffer a heart attack will compare to your previous calculations for Groups 1 and 2?

Objectives for the lesson

You will understand that:

☐ The change in a quantity can be expressed as an absolute change and a relative change.

☐ There is often ambiguity in the English language when talking about the change of a quantity that is represented by a percent (e.g., many rates).

You will be able to:

☐ Create graphs that show both absolute and relative changes in a rate (percent).

☐ Compute absolute and relative changes.

1)   Use estimation to complete the chart for the group of people described above.

Number of people who have heart attacks in 10 years without the drug	Number of people who have heart attacks in 10 years with the drug	Absolute change	Relative change

2) Based on your calculations in this lesson and in the previous lesson, critique the statement:

"A drug which reduces the risk of heart attack by 50% will most likely save many lives."

Drugs can help people lead healthier, more comfortable, and longer lives. They can also harm people. All drugs have side effects. It is important to consider carefully before deciding to take a medication. Consider this situation: Lipitor is a drug prescribed to reduce the risk of heart attack. It works by reducing the level of cholesterol in the blood. High cholesterol is believed to increase the risk of heart attack. The company that makes Lipitor published the following information:

Along with diet, Lipitor has been shown to lower bad cholesterol 39% to 60%.[20]

3) What information would you want to know if you were considering taking Lipitor?

---

[20] Go Low Cholesterol. (2011). Lipitor scare tactics. Retrieved March 19, 2013 from http://golowcholesterol.com/2011/08/04/lipitor-scare-tactics.

Lesson 9, Part A, Comparing categorical data
Theme: Risk Assessment

Grace Health Care Systems has received a grant to allow participation in the *100 Thousand Lives Campaign*. The grant affords the implementation of only one bundle.

An implementation team has been chosen to decide which bundle to adopt with the limited grant dollars. The team will use data to guide their decision.

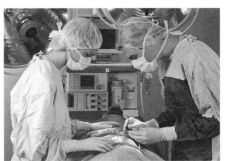

Credit: Tyler Olson/Fotolia

The team is deciding between the ventilator bundle, which helps prevent hospital-acquired infections from ventilator treatment, or the central line bundle, which helps prevent hospital-acquired infections from a central line.

Objectives for the lesson

You will understand that:

☐ Percentages involve a numerator (comparison value) and a denominator (base value).

You will be able to:

☐ Correctly identify the quantities involved in a verbal statement about percentages based on specific data.

☐ Calculate percentages and convert from a decimal representation to a percent.

☐ Read and use information presented in a two-way table.

Amy and Trevin are on the implementation team and are using the data summarized in the table below to guide their decision. The table excludes patients who received both treatments, so all patients in the table received only one treatment—central line or ventilator—not both. Take a moment to acquaint yourself with the table.

	Central Line Patients	Patients on Ventilator	Total
Developed Hospital-Acquired Infection (HAI)	20	22	
No HAI			158
Total	82	118	

1)   Complete the blank cells in the table.

Looking at the table, Amy says, "The hospital should implement the ventilator bundle, since more than 50% of the hospital-acquired infections involve patients on a ventilator." Trevin disagrees. He says, "Preventing hospital-acquired infections from central line infections is more important because nearly 24% percent of central line patients developed a hospital-acquired infection but only 19% of ventilator patients developed a hospital-acquired infection."

2) Read Amy's statement carefully. Is Amy correct? Justify your answer with calculations. Use the terms **comparison value** and **base value** in your explanation.

3) Now read Trevin's statement carefully. Is Trevin correct? Justify your answer with calculations. Use the terms **comparison value** and **base value** in your explanation.

4) Which bundle of interventions would you implement? Justify your answer based on the data.

Lesson 9, Part B, Interpreting percentages

Theme: Risk Assessment

As you saw in the previous lesson, statements about percentages are often confusing. Recall that HAI stands for "hospital-acquired infection," and consider the following two quantities:

Credit: Kot63/Fotolia

- Quantity 1 (Q_1): The percent of ventilator patients who developed a hospital-acquired infection.

- Quantity 2 (Q_2): The percent of HAI patients who are ventilator patients.

1)   Is Q_1 the same as Q_2? Explain your reasoning.

Objectives for the lesson

You will understand that:

☐ Percentages involve a numerator (comparison value) and a denominator (base value).

You will be able to:

☐ Correctly identify the quantities involved in a general verbal statement about percentages.

2)   Which quantity does the pie chart represent, Q_1 or Q_2? Explain your reasoning.

Ventilator Patients

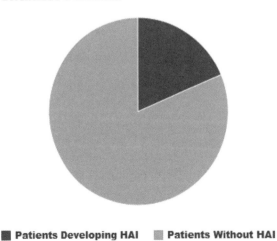

■ Patients Developing HAI   ■ Patients Without HAI

3) Label the pie chart below to represent the quantity you did not choose in question 2. Include a title naming the group representing the whole "pie" and labels for each group represented by a section of the pie.

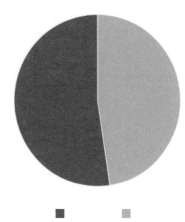

4) The Centers for Disease Control and Prevention (CDC) estimates that 4.5% of hospital admissions in America develop a hospital-acquired infection (HAI).[21] *Hospitals & Health Networks Daily* reports that almost one-third of these infections derive from the insertion of a central line.[22] Moreover, the goal within the health care system is to reduce infections due to central line insertion by 50%.

Part A: Approximately how many hospital admissions develop a hospital-acquired infection out of every 1,000 U.S. hospital admissions?

Part B: Approximately, how many hospital admissions develop a hospital-acquired infection out of every 1,000,000 U.S. hospital admissions?

Part C: Approximately, how many hospital admissions (out of every 1,000,000 U.S. admissions) lead to cases of HAI attributed to central line insertions?

---

[21] Douglas, II, R. D. (2009). The direct medical costs of healthcare-associated infections in U.S. hospitals and the benefits of prevention. Retrieved April 19, 2014, from http://www.cdc.gov/HAI/pdfs/hai/Scott_CostPaper.pdf

[22] Watson, R. (2012). A campaign to eliminate CLABSIs. *H&HN Daily*. Retrieved June 10, 2014, from http://www.hhnmag.com/display/HHN-news-article.dhtml?dcrPath=/templatedata/HF_Common/NewsArticle/data/HHN/Daily/2012/Sep/watson092012-9260001357

Part D: Assume the health care system meets its goal of reducing HAI due to central line insertions by 50%. Making this assumption, approximately how many hospital admissions (out of every 1,000,000 U.S. admissions) will lead to cases of HAI attributed to central line insertions?

Part E: Suppose a particular hospital with 10,000 annual admissions has HAI incident rates on par with the CDC estimates. How many annual incidences of central line infections should the hospital *not exceed* to meet the goal of reducing these infections by 50%?

Lesson 9, Part C,
Do you trust the test?   Theme: Risk Assessment

A Streptococcal bacterial infection causes a severe sore throat known as strep throat, and it is cured using a particular course of antibiotics. However, a Staphylococcus bacterial infection also causes a sore throat, and it is treated using a different course of antibiotics. Therefore, correctly identifying the cause of a sore throat is crucial to prescribing the correct treatment.

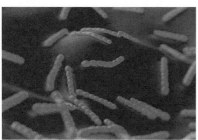

Credit: royaltystockphoto/Fotolia

Have you ever contracted strep throat? How did you know it was strep throat?

Objectives for the lesson

You will understand that:

☐ A percent can be used to express the likelihood (or probability) of a certain event.

☐ Selecting the appropriate comparison value and base value is crucial for calculating a percent correctly.

You will be able to:

☐ Extract relevant information from a two-way table.

☐ Select the appropriate values to compute probabilities.

Suppose 500 patients visit a walk-in clinic complaining of a sore throat and other symptoms consistent with strep throat. Clinic physicians order a diagnostic test. A positive test indicates that the patient has strep throat. A negative test indicates the patient does not have strep throat. However, medical tests are not 100% accurate, so the results of the test may be wrong. This means the test may be mistaken even when carefully administered.

1)   Complete the table to show how often the test correctly diagnosed cases of strep throat. Remember that a positive test indicates the patient has strep throat but that the test is not always correct.

Causes of Sore Throat in Patients

	Sore Throat Patients with Streptococcal Infection	Sore Throat Patients with Other Infections/Causes	Total
Positive test result	96		
Negative test result	24		385
Total		380	500

Use the figures or numbers in the table to answer the questions. Report probabilities using percent (%).

2) The **sensitivity** of a medical test equals the probability the test is positive when it should be positive. Compute the sensitivity of the test for strep throat using the steps below.

Part A: How many patients have a Streptococcal infection?

Part B: How many of the patients with a Streptococcal infection received a positive test?

Part C: If a patient has a Streptococcal infection, what is the probability that the test will correctly diagnose the patient with a positive result? The answer is the **sensitivity** of the test.

3) The **specificity** of a medical test equals the probability the test is negative when it should be negative. Compute the specificity of the test for strep throat using the steps below.

Part A: How many patients do not have a Streptococcal infection?

Part B: How many of the patients without a Streptococcal infection received a negative test?

Part C: If a patient does not have a Streptococcal infection, what is the probability that the test will correctly result in a negative outcome? The answer is the **specificity** of the test.

Lesson 9, Part D,
Do you trust the test? (continued)   Theme: Risk Assessment

Doctors diagnose symptoms, but they also rely on medical tests to confirm their diagnosis. Unfortunately, medical tests involve some chance of error. In some cases, patients test positive for a disease that they do not have. In other cases, the patient has the disease but tests negative. In this lesson, you will evaluate the accuracy of the test from the previous lesson.

Credit: Tatiana Shepeleva/Fotolia

Objectives for the lesson

You will understand that:

☐ A percent can be used to express the likelihood (or probability) of a certain event.

☐ Selecting the correct comparison value and base value is important in calculating percentages.

☐ Evaluating the accuracy of a test can depend on what data you focus on.

You will be able to:

☐ Extract relevant information from a two-way table.

☐ Select the appropriate values to calculate probabilities.

☐ Calculate and interpret probabilities of errors: percent of positive results that are false positives and percent negative results that are false negatives.

Refer to this table from Lesson 9, Part C.

Causes of Sore Throat in Patients

	Sore Throat Patients with Streptococcal Infection	Sore Throat Patients with Other Infections/Causes	Total
Positive test result	96	19	115
Negative test result	24	361	385
Total	120	380	500

1)   Think about the patients with a Streptococcal infection who received a negative test result. These test results are **false negatives**. How many false negatives appear in the table?

2)   A patient receives a negative result on the test for strep throat. What is the chance— rounded to the nearest tenth of a percent—that this negative result is a false negative?

3)   What do you think a **false positive test result** means? How many patients in the table received a false positive result?

4)   A doctor finds that a patient gets a positive result on the test.

   Part A: Knowing that the patient's test is positive, what is the approximate chance the result is a false positive?

   Part B: How should the doctor think about this percentage? What should the doctor do with this information?

5)   You can use different percentages to describe the accuracy of a test. A test is **accurate** when it results in a low percent of errors (**false positives** and **false negatives**). Pick one figure or percentage that you think best describes the accuracy of the test. Explain what this figure says about the test and why you picked this figure.

6)   Now, think about how to use a figure or percentage to describe the degree to which the test is **inaccurate**. A test is inaccurate when it results in high rates of errors (**false positives** and **false negatives**). Pick one figure to describe the inaccuracy of the test. Explain what this figure says about the test and why you picked this figure.

Lesson 10, Part A, Population density   Theme: Civic Life

In your preview assignment, you learned and worked with the definition of population density.

1) What would we need to know in order to calculate the population density of this class?

Credit: blvdone/Fotolia

Objectives for the lesson

You will understand that:

☐ Population density is a ratio of the number of people to the area of the region and its unit is people per unit of area.

You will be able to:

☐ Calculate a unit rate.

☐ Compare and contrast populations using their population densities.

2) Record the results of the class demonstrations and discussions in your notes.

How crowded is China compared to the United States? In 2010, in the United States, approximately 309,975,000 people occupied 3,717,000 square miles of land. In China, approximately 1,339,190,000 people lived on 3,705,000 square miles of land. Use this information to answer the following questions.

3) A student calculates the population densities of China and the United States. He decides that China is less dense than the United States. Using your estimation skills, decide if you think this student's calculation is correct.

4) Calculate the densities (per square mile) of the Chinese and U.S. populations. Based on your calculation, how many times more dense is the more crowded population? Be ready to share your calculations during the class discussion.

Lesson 10, Part B, Density proportions   Theme: Civic Life

Recall the density proportion in the previous activity of:

$$\frac{1 \text{ person}}{4 \text{ square feet}} = \frac{0.25 \text{ person}}{1 \text{ square feet}}$$

1) What would the population density be if one billion people each stood on adjacent 2-foot by 2-foot squares?

Credit: Rawpixel/Fotolia

Objectives for the lesson

You will understand that:

- ☐ **Population density** is a ratio of the number of people to the area of the region.
- ☐ Population density may be described proportionately to compare populations.
- ☐ **Proportional** means an increase or decrease based on a constant ratio.

You will be able to:

- ☐ Calculate population densities.
- ☐ Calculate population density proportions from density ratios.
- ☐ Solve a proportion by first finding a unit rate and then multiplying appropriately.

2) Now picture the billion people, each standing on the adjacent 2' x 2' squares. Calculate the population density per square mile. Be ready to explain your reasoning after working with your group members. (Recall that 1 mile = 5,280 feet.)

3) In your preview assignment, you calculated several population densities. If this campus had the same population density as Alaska, how many people would be on campus?

4) How many people would be on campus if the population density were equal to that of New Jersey?

5) Most of the world outside the United States uses the metric system of measurement, so it is often useful to make comparisons between the American system and the metric system. Bangladesh has a population density of 1,127 people per square kilometer.[23] (Note: 1 kilometer = 0.62 mile.) If you convert the density of Bangladesh to people per square mile, will the numerical value be larger or smaller than 1,127? Explain your reasoning.

---

[23] List of sovereign states and dependent territories by population density. In *Wikipedia*. Retrieved May 30, 2014, from http://en.wikipedia.org/wiki/List_of_sovereign_states_and_dependent_territories_by_population_density.

Lesson 10, Part C, State population densities   Theme: Civic Life

The approximate area of our state is _____.

The estimated current population of our state is

_____.

Credit: iQoncept/Fotolia

1) Estimate (no calculator!) the current population density of our state. Be ready to explain your strategy.

Objectives for the lesson

You will understand that:

☐ Population density is a ratio.

You will be able to:

☐ Estimate between which two powers of 10 a quotient of large numbers lies.

☐ Calculate a unit rate.

2) Now you will complete the comparison table you began in your notes from the preview assignment by estimating the density of each state. (The complete state list is provided after question 4. The densities you calculated exactly in the preview questions in your assignment are already filled in for you.)

Density < 10 people/mi^2	10–100 people/mi^2	100–1,000 people/mi^2	Density > 1,000 people/mi^2
Alaska			

3) Write a few sentences about the densities of the states with the greatest and lowest population density.

4) Did you use one or several estimation strategies while completing the table? Explain.

State	Land Area (Square Miles)	2010 Population[24]	Estimated Population Density (People/mi^2)	
Alabama	50,744	4,779,736		
Alaska	571,951	710,231	(Calculated)	1.2
Arizona	113,635	6,392,017		
Arkansas	52,068	2,915,918		
California	155,959	37,253,956		
Colorado	103,718	5,029,196		
Connecticut	4,845	3,574,097		
Delaware	1,954	900,877		
District of Columbia	61	601,723		
Florida	53,927	18,801,310		
Georgia	57,906	9,687,653		
Hawaii	6,423	1,360,301		
Idaho	82,747	1,567,582	(Calculated)	18.9
Illinois	55,584	12,830,632		
Indiana	35,867	6,483,802		
Iowa	55,869	3,046,355		
Kansas	81,815	2,853,118		
Kentucky	39,728	4,339,367	(Calculated)	109.2
Louisiana	43,562	4,533,372	(Calculated)	104.1
Maine	30,862	1,328,361		
Maryland	9,774	5,773,552		
Massachusetts	7,840	6,547,629		
Michigan	56,804	9,883,640		
Minnesota	79,610	5,303,925		

---

[24] Retrieved May 8, 2011 from http://en.wikipedia.org/wiki/U.S._state#List_of_states.

State	Land Area (Square Miles)	2010 Population[25]	Estimated Population Density (People/mi^2)	
Mississippi	46,907	2,967,297		
Missouri	68,886	5,988,927		
Montana	145,552	989,415		
Nebraska	76,872	1,826,341	(Calculated)	23.8
Nevada	109,826	2,700,551		
New Hampshire	8,968	1,316,470	(Calculated)	146.8
New Jersey	7,417	8,791,894	(Calculated)	1185.3
New Mexico	121,356	2,059,179		
New York	47,214	19,378,102		
North Carolina	48,711	9,535,483		
North Dakota	68,976	672,591		
Ohio	40,948	11,536,504		
Oklahoma	68,667	3,751,351		
Oregon	95,997	3,831,074		
Pennsylvania	44,817	12,702,379		
Rhode Island	1,045	1,052,567		
South Carolina	30,109	4,625,364		
South Dakota	75,885	814,180	(Calculated)	10.7
Tennessee	41,217	6,346,105		
Texas	261,797	25,145,561		
Utah	82,144	2,763,885		
Vermont	9,250	625,741		
Virginia	39,594	8,001,024		
Washington	66,544	6,724,540	(Calculated)	101.1
West Virginia	24,078	1,852,994		
Wisconsin	54,310	5,686,986	(Calculated)	104.7
Wyoming	97,100	563,626		
50 states + DC	**3,537,438**	**308,745,538**		

[25] Wikipedia. U.S. state. Retrieved May 8, 2011, from http://en.wikipedia.org/wiki/U.S._state#List_of_states.

| Lesson 10, Part D, Apportionment | Theme: Civic Life ANSWERS |

The United States Census is conducted every 10 years. The U.S. Census Bureau counts the population and collects other information.

Credit: Jcamilobernal/Fotolia

1)   What are some reasons why the government wants this information?

Objectives for the lesson

You will understand that:

☐ A relative change is different from an absolute change.

☐ A relative change is always a comparison of two numbers.

You will be able to:

☐ Calculate a relative change.

☐ Explain the difference between a relative change and absolute change.

2)   The 2000 and 2010 population counts for the states in the New England region of the United States are given in the table. Calculate the changes in population.

Change in New England States' Population from 2000 to 2010

New England Region	2000 Population	2010 Population	Absolute Change	Relative Change
Maine	1,274,923	1,328,361		
Massachusetts	6,349,097	6,547,629		
New Hampshire	1,235,786	1,316,470		
Rhode Island	1,048,319	1,052,567		
Vermont	608,827	625,741		

3)   What is the overall relative change in the New England region?

4)   Which state in this table grew the most? (Hint: There could be more than one answer, so explain your reasoning!)

5)   Each state sends representatives to the U.S. Congress based on its population. Currently, the law mandates a total of 435 representatives. After the 2010 census, Massachusetts actually lost one seat in the House. Can you think of an explanation for this?

Change in the Number of Representatives by State from 2000 to 2010[26]

State	Gain
Texas	4
Florida	2
Arizona	1
Georgia	1
Nevada	1
South Carolina	1
Utah	1
Washington	1

State	Loss
New York	2
Ohio	2
Illinois	1
Iowa	1
Louisiana	1
Massachusetts	1
Michigan	1
Missouri	1
New Jersey	1
Pennsylvania	1

6)   What do you think happened to the representation of the other states not listed above?

---

[26] U.S. Census Bureau. (2011). Congressional apportionment. Retrieved May 30, 2014, from www.census.gov/prod/cen2010/briefs/c2010br-08.pdf.

7)   Most of the states in the "loss" table actually gained population but are growing at a slower rate than the states in the "gain" table. Michigan is the only state that actually lost population during that period, from 9,938,444 in 2000 to 9,833,640 in 2010. Determine the absolute and relative change in Michigan's population and write a contextual sentence about your results.

Lesson 11, Part A, Formulating a plan   Theme: Personal Finance

Bob and Carol Mazursky have recently purchased their first home. The scale model shows the rectangular lot, the house, the driveway, and the backyard.

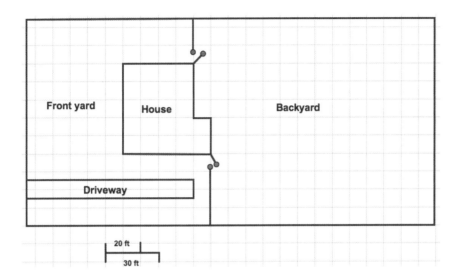

1)   Determine two things you could figure out about the property based on the scale.

Objectives for the lesson

You will understand that:

☐ A variable can be used to represent an unknown quantity.

☐ Using a formula requires knowing what each variable represents.

☐ The units are helpful when working with formulas.

You will be able to:

☐ Use formulas from geometry.

☐ Evaluate an expression.

The Mazurskys are expecting their first child in several months and want to get the backyard fertilized and reseeded before the baby arrives.

2)   How does the information you recorded in question 1 relate to the ideas of fertilizing and reseeding, or what additional information do you need to determine?

Lesson 11, Part B, The costs of geometry   Theme: Personal Finance

The Mazurskys decide to begin their improvements with fertilizing and reseeding the backyard. They found an ad for Gerry's Green Team lawn service. Gerry came to their house and said that the job would take about four hours and would cost about $600.

Gerry's Green Team

Itemized Costs:

Grass seed	4 pounds per 1,000 sq. ft. @ $1.25 per pound
Fertilizer	50 pounds per 12,000 sq. ft. @ $0.50 per pound
Labor	$45 per hour

advertisement

Objectives for the lesson

You will understand that:

☐ The units are helpful when working with formulas.

You will be able to:

☐ Perform calculations that involve rates and measures to support financial decisions.

☐ Solve complex problems requiring multiple pieces of information and steps.

☐ Evaluate an expression.

1)   Is Gerry's estimate consistent with his advertisement?

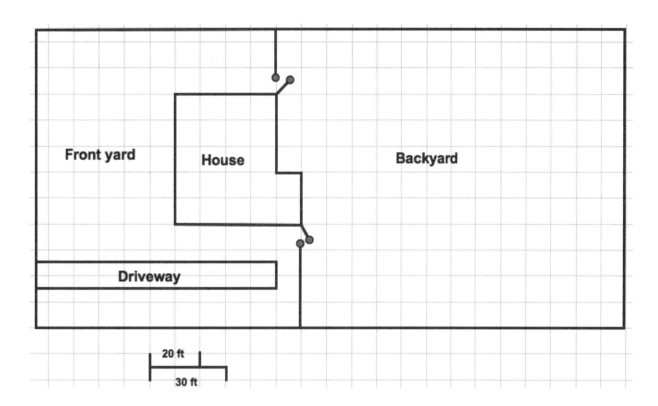

In the future, Bob and Carol plan to build a 48-inch-tall chain-link fence around the backyard with two gates on either side of the house. You will help them with this project in Practice Assignment 11.AB.

Lesson 11, Part C, Modifying and combining formulas

Theme: Personal Finance

The Mazurskys are doing more work in the backyard. There is a brick grill, and Bob and Carol want to make a concrete patio in the shape of a semicircle next to the grill. The concrete slab needs to be at least 2 inches thick. They will use 40-pound bags of premixed concrete. Each 40-pound bag makes 0.30 cubic feet of concrete and costs $6.50.

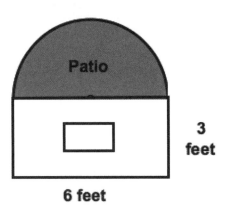

3 feet

6 feet

Objectives for the lesson

You will understand that:

- ☐ You can find formulas through the web and reference books.
- ☐ The units are helpful when working with formulas.

You will be able to:

- ☐ Use formulas from geometry and perform calculations that involve rates and measures to support financial decisions.
- ☐ Evaluate an expression.

1)   How much will the materials cost, including the 7.5% sales tax?

In the future, Bob and Carol plan to add a concrete patio on the side of the house adjacent to the driveway and new sod between the house and driveway. (Sod is a layer of grass that can be laid on top of the ground so the grass grows quickly.) You will help them with these projects in Practice Assignment 11.C.

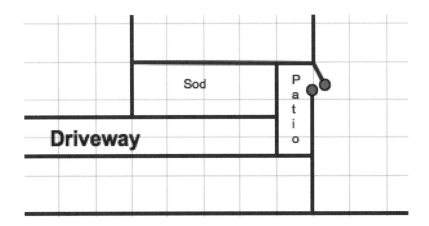

Lesson 12, Part A, Texting distance   Theme: Risk Assessment

Many cities and states have banned texting while driving because it is dangerous, but many people think that texting for a few seconds is not harmful. Suppose you are driving 45 miles per hour and you take your eyes off the road to answer a text.

Credit: Karen Roach/Fotolia

1) Predict the number of feet your car will travel during the 4 seconds it takes to reply.

Objectives for the lesson

You will understand that:

☐ Units provide meaning to the numbers you get from calculations.

☐ The units found in a solution may be used as a guide for the needed conversions.

You will be able to:

☐ Use units to determine which conversion factors are needed for dimensional analysis.

☐ Use a conversion factor to convert a rate.

2) To check your prediction, begin by setting up only the units (no numbers) that are needed to convert from miles per hour to feet per second. Refer to Resource **Dimensional Analysis**, if needed.

3) One student set up the unit calculation as shown. Decide if the problem is set up correctly. If not, correct it.

$$\frac{\text{miles}}{\text{hour}} \bullet \frac{\text{minutes}}{\text{hour}} \bullet \frac{\text{minutes}}{\text{second}} \bullet \frac{\text{miles}}{\text{feet}} = \frac{\text{feet}}{\text{second}}$$

4) Now place the numbers into your set-up and simplify. What is your speed in feet per second? How many feet will you travel in 4 seconds if you are driving 45 miles per hour? How close was your prediction?

5) Suppose that most text responses take between 2 and 6 seconds. Incorporate this information into your original dimensional analysis strategy and adjust your answer.

Lesson 12, Part B, The cost of driving   Theme: Personal Finance

Jenna's job requires her to travel. She can drive her own 2009 Toyota 4Runner, or she can rent a car. Either way, her employer will reimburse her for mileage.

Credit: ellisia/Fotolia

1) Should Jenna drive her own car or rent one?

     Part A: What do you need to know to calculate the cost of Jenna's driving her own car?

     Part B: What do you need to know to calculate the cost of Jenna's renting a car?

Objectives for the lesson

You will understand that:

- ☐ Units provide meaning to the numbers you get from calculations.
- ☐ The units found in a solution may be used as a guide to the operations required.

You will be able to:

- ☐ Write a rate as a fraction.
- ☐ Use a unit factor to convert a rate.
- ☐ Use dimensional analysis to help determine the factors in a series of operations to obtain an equivalent measure.

2) Jenna's 4Runner is expected to get 22 miles/gallon on the highway, or she can rent a Hyundai Elantra, which gets 40 miles/gallon.[27] On her next trip, she will drive 150 miles, and gas is expected to cost $3.59/gallon.

     Part A: Use your estimation skills to estimate which vehicle would cost more in gas. About how much more? Explain your strategy.

     Part B: What is the actual cost of the gas for each vehicle? Explain your strategy.

3) What is the total cost of the Elantra if rental is $98.98 plus 15.3% tax? (Gas is not included in this price, and the car must be returned to the agency with a full tank.)

---

[27] Source: www.fueleconomy.gov.

4)   Jenna's employer will reimburse her for the mileage using the rate set by the Internal Revenue Service. In 2013, that rate was 56.5 cents/mile.[28] How much "profit" will she make under each option?

5)   Are there other expenses or issues that might impact Jenna's decision?

_____

[28] Internal Revenue Service (2013). *Standard mileage rates for 2013*. Retrieved June 2, 2014, from http://www.irs.gov/uac/ Newsroom/2013-Standard-Mileage-Rates-Up-1-Cent-per-Mile-for-Business,-Medical-and-Moving.

Lesson 12, Part C,   Theme: Personal Finance
The true cost of driving   ANSWERS

In Part B, you found that the true cost of renting a car meant that better gas mileage doesn't tell the whole story.

Credit: Pictures news/Fotolia

1) What other expenses can you think of that impact Jenna's decision in addition to the cost of fuel and the fuel efficiency of the car?

Objectives for the lesson

You will understand that:

☐ Units can be used as a guide to set up calculations.

☐ Rounding can produce large differences in results, depending on the size of the numbers and the precision of the original values.

You will be able to:

☐ Solve a complex problem with multiple pieces of information and steps.

2) Calculate the true cost of the driving the 4Runner using these assumptions:

- **Insurance, registration, taxes:** Jenna spends $2,000 per year on these expenses, and last year she drove about 21,600 miles. (Note that these costs can vary greatly depending on location, vehicle, and the individual.)

- **Oil and other regular maintenance**: $40 every 3,000 miles.

- **Tires**: Tires for Jenna's car cost $920 and should be replaced every 50,000 miles.

- **Repairs**: In 2013, the website Edmonds.com estimated that yearly repairs on a 2009 4Runner would be approximately $577 based on 15,000 miles.[29]

- **Gas**: Jenna's 4Runner is expected to get 22 miles/gallon on the highway.

_____

[29] Edmunds. (n.d.) 2009 Toyota 4Runner True cost to own. Retrieved June 2, 2014, from http://www.edmunds.com/toyota/4runner/2009/tco.html?style=101056520.

3)   Which of the above expenses does Jenna have to pay, regardless of mileage?

4)   Since Jenna's trips vary in length, it may be useful for her to compare the cost per mile for renting to the cost per mile for driving her own car. What are these rates and what do they depend on?

Lesson 12, Part D
Can the true cost vary?
Theme: Personal Finance

When Jenna took a 150-mile trip, you calculated that driving her own 4Runner would be the least expensive route and give her more money left over from her reimbursement.

Credit: Hayati Kayhan/Fotolia

1) Do you believe the 4Runner would <u>always</u> be the best option? Why or why not? How could you check?

Objectives for the lesson

You will understand that:

☐ Units can be used as a guide to set up calculations.

You will be able to:

☐ Solve a complex problem with multiple pieces of information and steps.

☐ Investigate how changing certain values can affect the result of a calculation.

2) What factors would affect the cost of the two options?

3) Investigate various trip lengths to test your hypothesis from question 1.

4) Write to Jenna, explaining how she can decide if it is better to drive her own car or to get a rental. Your explanation should include information about what factors affect the cost of driving and why.

Begin making notes on what you want to say. If you do not finish during class, bring your explanation to the next class meeting.

Lesson 13, Part A, Algebra reaction   Theme: Civic Life

Credit: Stefan Schurr/Shutterstock

Reaction time is "the interval of time between application of a stimulus and detection of a response."[30] Athletes are often concerned with their reaction time.

1)   What are some situations in everyday life where reaction time is important?

Objectives for the lesson

You will understand that:

☐ A variable is a symbol that is used to represent a quantity that can change.

You will be able to:

☐ Evaluate an expression.

Testing reaction time:

- Person A holds a ruler vertically.

- Person B positions their thumb and first finger about 1 inch apart and on either side of the bottom of the ruler.

- Person A drops the ruler without warning, and Person B attempts to catch it as quickly as possible.

- Record the distance from the bottom of the ruler to the nearest ½ inch. Run three trials with each hand.

---

[30] Source: The American Heritage® Dictionary of the English Language, Fourth Edition. (2000). Boston, MA: Houghton Mifflin Company.

2) In the formula given below, d is the average distance in feet and t is the time in seconds. Use the formula to calculate your reaction time.

$$t = \sqrt{\frac{d}{16\,{}^{ft}\!/_{s^2}}}$$

Trial	Distance (inches)
1	
2	
3	
4	
5	
6	
Avg.	

Lesson 13, Part B, Breaking down a formula   Theme: Civic Life

In an earlier lesson, you determined that when you are driving 45 mph, you will travel 264 feet in 4 seconds, or 66 ft/sec. What if you apply the brakes?

1)   Predict the braking distance, the distance the car will travel after the brakes have been applied.

Credit: Ion Chiosea/123RF

Objectives for the lesson

You will understand that:

☐   A variable is a symbol that is used to represent a quantity that can change.

☐   Some variables in a formula can be held fixed in order to analyze the effect that the change in one variable has on another.

You will be able to:

☐   Evaluate an expression.

The formula for the braking distance of a car is $d = \dfrac{V_0^2}{2g(f+G)}$, where

V_0 = initial velocity of the car in feet per second (the velocity of the car when the brakes were applied)

d = braking distance (feet)

G = roadway grade (percent written in decimal form)

f = coefficient of friction between the tires and the roadway ($0 < f < 1$)

(Note: Good tires on good pavement provide a coefficient of friction of about 0.8 to 0.85.)

Constant:

g = acceleration due to gravity ($\approx 32.2 ft/s^2$ or $\approx 9.8\,m/s^2$)

Since g is a constant, this formula has four variables. To understand the relationships between the variables, you will hold two of them fixed. That leaves you with two variables—one that will affect the other. Since you want to see how speed affects braking distance, you will hold the other two variables, f and G, fixed.

2)   Consider a situation in which the coefficient of friction is 0.8 and the roadway grade is 0.05. Write a simplified form of the formula using these values.

3)   Calculate the braking distance when the speed is 45 mph (66 ft/sec). How does the distance compare to the prediction you made in question 1?

4)   When the speed is 45 mph (66 ft/sec), calculate the braking distance when the roadway grade is 0.15 and 0.30.

Lesson 13, Part C, Analyzing change in variables

Theme: Civic Life

Credit: Ion Chiosea/123RF

At the beginning of Lesson 13, Part B, you predicted the stopping distance when the speed was 45 mph. Were you surprised, or was your prediction fairly accurate? What if the speed were different?

1) How could you explore the effect that different speeds have on braking distance? (Keep the friction and grade fixed at $f = 0.8$ and $G = 0.05$.)

Objectives for the lesson

You will understand that:

☐ A variable is a symbol that is used to represent a quantity that can change.

☐ Some variables in a formula can be held fixed in order to analyze the effect that the change in one variable has on another.

You will be able to:

☐ Evaluate a rational expression.

☐ Establish a strategy to explore the pattern of changes in one variable while holding other variables fixed.

2) Determine the stopping distance for several speeds. Document your work thoroughly.

3) Based on your results, what conclusions can you draw?

4) How confident are you in your conclusions?

5) Predict the braking distance if you double the speed. What if you triple the speed? What are some other questions you have about the relationship between speed and braking distance?

Lesson 13, Part D, Analyzing change in variables (continued)
Theme: Civic Life

Credit: bizoon/123RFo

At the beginning of Lesson 13, Part C, you were asked to to explore the effect that different speeds have on braking distance. Were you surprised with the results?

1) How could you explore the effect that different grades have on braking distance? (Keep the friction fixed at $f = 0.8$ and same speed V_0.)

Objectives for the lesson

You will understand that:

☐ A variable is a symbol that is used to represent a quantity that can change.

☐ Some variables in a formula can be held fixed in order to analyze the effect that the change in one variable has on another.

You will be able to:

☐ Informally describe the change in one variable as another variable changes.

2) Determine the stopping distance for several values of G. Document your work thoroughly.

3) Based on your results, what conclusions can you draw?

You have now used several different formulas in this course. In early lessons, you developed several spreadsheet formulas (e.g., multiplying a tax rate times the sum of several cells). In Lesson 11, you used geometric formulas for area and volume. In Lesson 12, you developed your own formula to analyze Jenna's driving costs.

In this lesson, you used a formula that was more complex and probably less familiar to you. Almost every field has specialized formulas, but they all depend on three basic skills:

- Understanding and knowing how to use variables, including the use of subscripts.

- Understanding and knowing how to use the order of operations.

- Understanding and knowing how to use units, including dimensional analysis.

With these three skills, you will be able to use formulas in any field.

Lesson 14, Part A, Body mass index   Theme: Risk Assessment

The body mass index (BMI) is a measure of body fat based on height and weight. It is a standard tool for helping judge the amount of body fat you have. Carrying excess body fat puts people at greater risk for health problems such as heart disease, cancer, diabetes, and stroke. BMI can be calculated using a simple ratio based on a person's height and weight. Your BMI is considered to be in the normal range if it is between 18.5 and 25.

1) BMI is considered a better predictor of health than weight alone. Jot down your ideas about why this statement would be true, then share with another student.

Credit: Andy Dea/Fotolia

Objectives for the lesson

You will understand that:

☐ Substituting for one variable can yield a simplified formula.

You will be able to:

☐ Explicitly write out the order of operations to evaluate a given formula.

BMI can be calculated with the following formula, where the weight is in pounds and the height is in inches.

$$BMI = \frac{Weight}{Height^2} \times 703$$

2) Joe is 5 feet, 10 inches tall. Substitute his height into the formula.

3) You have created a new formula that applies only to people who are Joe's height. What are the only variables that remain in your new formula?

4) Using your simplified formula, calculate Joe's BMI if he weighs 175 pounds. How does Joe's BMI change if he gains 10 pounds? If he loses 10 pounds?

5) Discuss with your group how you arrived at Joe's BMI value. For example, did you multiply, add, subtract, etc., and what did you do first? Outline the steps you took to calculate the BMI when given the height and weight. Be specific.

Lesson 14, Part B, Target weight   Theme: Risk Assessment

Have you ever seen a video running backward? What happens? For example, if the original video shows a person opening the door, getting into the car, and starting the car, what does the reversed video look like?

Credit: iQoncept/Shutterstock

Objectives for the lesson

You will understand that:

☐ Multiplication and division are inverse operations.

☐ Solving for a variable includes isolating it by "undoing" the operations in an equation.

You will be able to:

☐ Solve for a variable in an equation.

☐ Explicitly write out the order of operations to solve a given equation.

BMI can be calculated by the following formula, where the weight is in pounds and the height is in inches.

$$BMI = \frac{Weight}{Height^2} \times 703$$

Recall that a person is considered to be underweight if his or her BMI is less than 18.5. A person is considered to be overweight if BMI is between 25 and 30, and obese if BMI is over 30.

1) Phillip is 6 feet, 3 inches tall. What is Phillip's BMI formula? Determine how much he would weigh for each of the following BMI values. Record your steps for part (a) in words.

   Part A: 18.5

   Part B: 25

   Part C: 30

2) Sally's weight is 115 pounds. What is her simplified BMI formula? How does it differ from Phillip's?

3) Sally has hit a growth spurt! Four months ago, she was 5 feet, 3 inches tall. Now she is 5 feet, 6 inches tall. She has maintained her weight at 115 pounds. Find her BMI both four months ago and now. How has it changed?

4) Sally continues to grow taller while maintaining her weight. She eventually reaches an unhealthy BMI of 18. How does this problem differ from question 1, where you were given BMI values for Phillip?

Lesson 14 Part C, Blood alcohol content   Theme: Risk Assessment

Blood alcohol content (BAC) is a measurement of how much alcohol is in someone's blood. It is usually measured as a percentage, so a BAC of 0.3% is three-tenths of 1%. That is, there are 3 grams of alcohol for every 1,000 grams of blood. A BAC of 0.05% impairs reasoning and the ability to concentrate. A BAC of 0.30% can lead to a blackout, shortness of breath, and loss of bladder control. In most states, the legal limit for driving is a BAC of 0.08%.[31]

Credit: zstock/Fotolia

BAC is usually determined by a breathalyzer, urinalysis, or blood test. However, Swedish physician E.M.P. Widmark developed the following equation for estimating an individual's BAC. This formula is widely used by forensic scientists:[32]

$$B = -0.015 \cdot t + \frac{2.84 \cdot N}{W \cdot g}$$

Objectives for the lesson

You will understand that:

☐ The location of a variable is important to its effect on the size of an expression.

You will be able to:

☐ Explicitly write out order of operations to evaluate a given formula.

The variables in the formula are defined as:

B = percentage of BAC

N = number of "standard drinks" (N should be at least 1)

   (A standard drink is one 12-ounce beer, one 5-ounce glass of wine, or one 1.5-ounce shot of liquor.)

W = weight in pounds

g = gender constant, 0.68 for men and 0.55 for women

t = number of hours since the first drink

---

[31] Blood alcohol content. In *Wikipedia*. Retrieved June 7, 2014, from http://en.wikipedia.org/wiki/Blood_alcohol_content. Note that different countries measure BAC in different ways involving mass and volume.

[32] Gullberg, R. G. (2007, August). *Estimating the uncertainty associated with Widmark's equation as commonly applied in forensic toxicology*. Paper presented at the T2007 Conference. Retrieved June 7, 2014, from www.icadts2007.org/print/108widmarksequation.pdf.

1) Look at the right side of the equation. How do the variables and their location make sense in calculating BAC? For example, based on their locations, which variables will make BAC larger? Which will make it smaller?

2) Consider the case of a male student who has five beers and weighs 180 pounds. Simplify the equation as much as possible for this case. What variables are still unknown in the equation?

3) Using your simplified equation, find the estimated BAC for this student one, three, and five hours after his first drink. What patterns do you notice in the data?

4) Record the sequence of steps you took to get from "time" to BAC. Be specific. For example, did you multiply, add, subtract, etc.? What values and in what order?

Lesson 14 Part D, Balancing blood alcohol
Theme: Risk Assessment

Recall Widmark's blood alcohol formula:

$$B = -0.015 \cdot t + \frac{2.84 \cdot N}{W \cdot g}$$

The variables in the formula are defined as:

B = percentage of BAC

N = number of "standard drinks" (N should be at least 1)

   (A standard drink is one 12-ounce beer, one 5-ounce glass of wine, or one 1.5-ounce shot of liquor.)

W = weight in pounds

g = gender constant, 0.68 for men and 0.55 for women

t = number of hours since the first drink

Credit: zstock/Fotolia

Find the simplified formula for the 180-pound male who drank five beers. What sequence of steps did you take to determine his BAC after 3 hours? Remember to be specific about operations used, values, and order.

Objectives for the lesson

You will understand that:

☐ Addition/subtraction and multiplication/division are inverse operations.

☐ Solving for a variable includes isolating it by "undoing" the operations in an expression.

You will be able to:

☐ Solve for a variable in an equation.

☐ Explicitly write out the order of operations to solve a given equation.

1) How long will it take for this student's BAC to be 0.08, the legal limit? How long will it take for the alcohol to be completely metabolized, resulting in a BAC of 0.0?

2) A female student, weighing 110 pounds, plans on going home in two hours.

   Part A: Determine this student's simplified formula.

   Part B: If she has one glass of wine now, what will her BAC be when she leaves to go home?

   Part C: What if she has three glasses of wine?

   Part D: Calculate an estimate of how many drinks she can have and keep her BAC less than 0.08.

Lesson 15, Part A, Proportional reasoning in art

Theme: Civic Life

The Heart Health Association is holding a 5K Fun Run/Walk. An artist has donated his time to enlarge the logo for the advertising banners and the t-shirts that will be given to each runner.

Many professionals such as graphic artists, architects, and engineers work with objects that are enlarged or shrunk. In this lesson, you will explore the mathematics behind these changes in size.

Credit: wavebreakmedia/Shutterstock

Objectives for the lesson

You will understand that:

☐ Proportional relationships are based on a constant ratio.

You will be able to:

☐ Set up a proportion based on a contextual situation.

☐ Solve a proportion.

	A	B	C	D	E	F	G	H	I
1		Height (inches)	Width (inches)		Height (inches)	Width (inches)		Height (inches)	Width (inches)
2	Original	1.67	2.49		1.67	2.49		1.67	2.49
3	T-shirt	1.67	9.96		6.68	8.72		8.35	12.45
4	Banners	1.67	24.90		25.05	32.37		23.38	34.86

1) The artist wants to set up a spreadsheet to calculate the dimensions of the graphics for the t-shirt and banner so that she can reuse it for future projects. She is not sure how to write the formulas correctly and tries different options. Which of the three options creates a proportional relationship? (Hint: If a relationship is proportional, the image is not distorted.)

2) Use the correct option from question 1. Write the spreadsheet formulas that are used to calculate the width for each version (t-shirt and banner). Use the column and row labels shown above for the cell references. Round the ratio to the nearest hundredth.

3) You are a graphic artist hired to make a billboard for a college. The original college logo is $2\frac{1}{4}$ inches (width) by $3\frac{3}{8}$ inches (height). You need to enlarge it to a height of 6 feet. How wide will the enlarged version be?

4) Write a brief memo to the artist in question 1, explaining how she can tell which option is correct.

Lesson 15, Part B, Proportion solutions   Theme: Civic Life

In the previous activity, you began with a college logo that was 2.25" wide by 3.375" long and enlarged it so that the new length was 6 feet. You may have set up a proportion such as: Let x = width.

$$\frac{2.25}{3.375} = \frac{x}{6}$$

Credit: chromaco/Fotolia

1) Is the proportion below also a correct representation? If so, how would you solve it?

$$\frac{3.375}{2.25} = \frac{6}{x}$$

Objectives for the lesson

You will understand that:

☐ Proportional relationships are based on a constant ratio.

☐ Rules for solving equations can be applied to many types of equations in different forms.

You will be able to:

☐ Set up a proportion based on a contextual situation.

☐ Solve a proportion using algebraic methods.

Solve each equation. Round to the nearest tenth.

2) $\dfrac{12.7}{x} = \dfrac{0.2}{3}$

3) $\dfrac{8,500}{4,200} = \dfrac{x}{5}$

4) Many small engines for saws, motorcycles, and utility tractors require a mixture of oil and gas. If an engine requires 20 ounces of oil for 5 gallons of gas, how much oil would be needed for 8 gallons of gas?

5) Write your own problem, similar to question 4, about a situation that has meaning in your life. Your example should not use a unit rate (a comparison to 1 such as 35 miles per 1 gallon).

Lesson 15, Part C, Solving equations   Theme: Risk Assessment

Credit: Maridav/Shutterstock

Solving equations such as blood alcohol content (BAC) and proportional equations for resizing graphics is an important skill. Mathematical models are often constructed to represent real-life situations. Being able to use these equations fully includes being able to solve for unknown variables in the equation. Below are two scenarios for you to practice and enhance your equation-solving skills. With each problem, check that the answer is reasonable, given the context, and that you have included the correct units with your solution.

Objectives for the lesson

You will understand that:

☐ Many equations can be solved by following the basic rules of undoing and keeping the equation balanced.

You will be able to:

☐ Solve equations that require simplification before solving.

☐ Solve for a variable in terms of other variables.

1) Paula has two options for going to school. She can carpool with a friend or take the bus. Her friend estimates that driving will cost 22 cents per mile for gas and 8.2 cents per mile for maintenance of the car. Additionally, there is a $25 parking fee per week at the college. If Paula carpools, she would pay half of these costs. The cost of the carpool can be modeled by the following equation, where C is the cost of carpooling per week and m is the total miles driven to school each week:

$$C = \frac{1}{2}(0.082m + 0.22m + 25)$$

Part A: Explain what each term in the equation represents.

Part B: Find the total weekly carpooling cost if the commute to school is 7 miles each way and Paula goes to school three times a week.

Part C: A weekly bus pass costs $22.00 dollars. How many total miles must Paula commute to school each week for the carpool cost to be equal to the bus pass? How many trips to school each week must Paula make for the bus pass to be less expensive than carpooling?

2)   Recall Widmark's equation for BAC. In the case of the average male who weighs 190 pounds, you can simplify Widmark's formula to get

$$B = -0.015t + 0.022N$$

Forensic scientists often use this equation at the time of an accident to determine how many drinks someone had. In these cases, time (t) and BAC (B) are known from the police report. The crime lab uses this equation to estimate the number of drinks (N).

Part A: Find the number of drinks if the BAC is 0.09 and the time is 2 hours.

Part B: Since the lab uses the formula to solve for N over and over, it is easier if the formula is rewritten so that it is solved for N. In other words, N is isolated on one side of the equation and all other terms are on the other side. Solve for N in terms of t and B.

Part C: Use the new formula to find the number of drinks if the BAC is 0.17 and the time is 1.5 hours.

3)   Solve $a + 2b = 12$ for b.

4)   Solve $y = \dfrac{-2}{3x - 1}$ for x.

5)   Solve $\dfrac{B}{D - A} = C$ for B.

Lesson 15, Part D, More work with equations

This lesson is different from others in this course. It does not include a problem-solving context. It focuses on building your confidence and skills in solving equations. Your instructor will provide you with equations to solve. After you have worked on these equations, answer the questions below.

Credit: Karen Keczmerski/123RF

Objectives for the lesson

You will understand that:

☐ Rules for solving equations can be applied in unfamiliar situations.

You will be able to:

☐ Solve equations in a variety of forms.

☐ Recognize when you do not have enough tools to solve an equation.

1)   Solve the equations given to you by your instructor.

2)   Think back on the different types of equations you have solved in this course. What rules can be applied to all equations?

3)   Apply these rules to the equation below. Note what you can and cannot do in this situation.

$$2x^2 + 5x - 8 = 12$$

Lesson 15, Part E, Proportional viewing   Theme: Civic Life

Credit: karandaev/Fotolia

You have probably watched movies on TV in a "letterbox" format in which there is a dark band above and below the image. This format is used to compensate for the difference between the dimensions of a movie screen compared to a TV screen. The following information is excerpted from www.widescreen.org.[33]

> Since 1955, most movies were (and are) filmed in a process where the width of the visual frame is between 1.85 to 2.4 times greater than the height. This means that for every inch of visual height, the frame as projected on the screen is between 1.85 to 2.4 times as wide.

The result is a panoramic view that, when used properly, can add a greater breadth and perception of the environment and mood of a movie.

This formula is called an **aspect ratio**. The aspect ratio is usually stated using the colon form. For example:

o   A movie that is 1.85 times wider than it is high has an aspect ratio of 1.85:1.

o   Similarly, a movie that is 2.35 times wider than it is high has an aspect ratio of 2.35:1.

Modern televisions come in two aspect ratios—1.33:1 (or 4:3), which has been the standard since television became popular, or 1.77:1 (more commonly known as 16:9), which is quickly becoming the new standard. However, neither of these aspect ratios is as wide as the vast majority of modern movies, most of which are either 1.85:1 or 2.35:1.

> "When you watch a movie on your television screen, you're not necessarily seeing it the way it was originally intended. As a director, when I set up a shot and say that there are two people in the frame, with the wide screen, I can hold both with one person on each end of the frame. When that shot is condensed to fit on your TV tube, you can't hold both [actors] . . . and the intent of the scene is sometimes changed as a result."

> —Leonard Nimoy,

> Commentary for the Director's Edition of *Star Trek IV: The Voyage Home*

---

[33] Widescreen.org. (n.d.) What is widescreen? Retrieved June 15, 2014, from http://www.widescreen.org/widescreen.shtml.

Objectives for the lesson

You will understand that:

☐ Proportional relationships are based on a constant ratio.

You will be able to:

☐ Set up a proportion based on a contextual situation.

1) Part A: Demonstrate mathematically that an aspect ratio of 2.35:1 for a movie is not proportional to the ratio of 4:3 for a TV. Provide written explanation as needed.

Part B: Explain why a picture with dimensions of 2.35:1 cannot be resized to have dimensions of 4:3 without changing the picture.

2) A standard HDTV screen with a width of 26 inches has a height of 19.5 inches. What is the ratio of width to height of the HDTV? Be sure to simplify your answer.

Lesson 16, Part A, Describing rates   Theme: Personal Finance

Interpreting numbers in advertising or product descriptions can be problematic. Data can be presented in ways that point to faulty conclusions. The table below presents the approximate distances that three vehicles can travel on a tank of gas.

Credit: Hayati Kayhan/Fotolia

Model	Miles per Tank
Suburban	651.0
Focus	446.4
Volt	380.0

1) Write down one conclusion or impression you can draw from the data in the table.

2) Compose two questions you would like to have answered to help you interpret the information in the table.

Objectives for the lesson

You will understand:
- ☐ The meaning of slope in a problem situation.
- ☐ How slope affects the behavior of a graph.

You will be able to:
- ☐ Calculate slope in direct variation situations.
- ☐ Express slope with appropriate units.
- ☐ Interpret slope in a problem situation.

3) A Suburban has a fuel capacity of 31 gallons. How can you use this new information to find how far a Suburban can travel on one gallon of gas?

4) Calculate the distance that a Suburban can travel on one gallon of gas. Write your solution as a unit fraction, with labels in both the numerator and denominator.

5) Write a contextual sentence interpreting the meaning of the number you found in question 4.

6) Create a table like the one shown. Use your answer to question 4 to calculate the distance a Suburban can travel on the given gasoline amounts.

x = Gallons of Gas	1	2	3	4	5	6	7
y = Distance (in miles)							

7) Notice that every time the amount of gas increases by one gallon, the distance increases by a fixed amount. This fixed amount represents the **slope** in the relationship between gas consumption and distance. What is the fixed amount?

8) Create a coordinate plane like the one shown below. Plot the coordinates of the points that represent (gas, distance) for the Suburban.

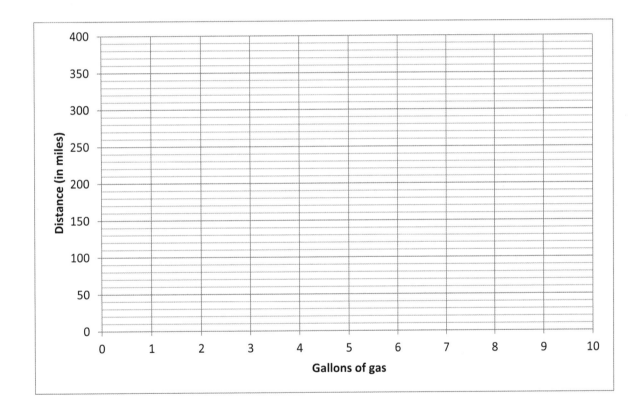

9) Write a few sentences in your own words, describing how the "behavior" of your graph relates to the information in the problem.

Lesson 16, Part B, Comparing rates   Theme: Personal Finance

In Part A, given some additional information, you were able to convert what you knew about "miles per tank" into information about "miles per gallon" for the Suburban. You also created a graph.

Credit: mrallen/Fotolia

Model	Miles per Tank
Suburban	651.0
Focus	446.4
Volt	380.0

1) What do you think the graph of (gas, distance) will look like for the Focus and for the Volt? (Remember that the independent values on the horizontal axis are gallons, and the dependent values on the vertical axis are miles.)

2) What information do you need so that you can check your prediction?

Objectives for the lesson

You will understand:

☐ How slope affects the behavior of a graph.

You will be able to:

☐ Calculate slopes in direct variation situations.

☐ Express slope with appropriate units.

☐ Compare slopes in a problem situation.

3) A Ford Focus has a fuel capacity of 12.4 gallons, while the capacity of a Chevy Volt is 9.3 gallons. Use this new information to calculate the slope for each linear relationship between the gallons of gas each car consumes and the average distance it can travel. Write your answers as unit rates, with units labeled in both the numerator and denominator.

4) Create tables of values for the Focus and for the Volt, as you did in Lesson 16, Part A. Then plot your values on the same coordinate plane as the Suburban information. Find a way to clearly distinguish the three graphs on the same plane.

5) How do the three slopes you found, and the graph you created, allow you to compare the fuel cost of operating the three cars in a meaningful way?

6) Give a reason why someone might buy each car based on the information in these lessons.

Lesson 16, Part C, Interpreting change   Theme: Personal Finance

The Smith family is thinking about changing cell phone carriers and possibly getting additional phones. After doing some research, they found a plan that includes talk, text, and 1,000 minutes that can be shared among the users of all of the phones.[34]

Number of phones	2	3	6
Cost per month ($)	99.99	109.99	139.99

Credit: kalim/Fotolia

1) Look at the first two data points in the table. What do they tell you about the slope of the relationship between the number of phones and the monthly cost? Write a sentence about this relationship. Be specific and be sure to include units!

Objectives for the lesson

You will understand:

☐ The meaning of slope in problem situations.

☐ Any two points on a line can be used to find the slope.

You will be able to:

☐ Calculate slope from two data points in a linear relationship.

2) Now refer to the second and third data points from the table. How much more will it cost to have six phones rather than three? Describe how you arrived at this conclusion.

3) How does your response to question 2 compare to question 1?

4) Plan how you would find the slope if you were using the first and third data points. You can describe your plan in words or set up a mathematical expression to represent the arithmetic.

5) Now use your plan from question 4 to calculate the slope using the first and third points.

6) Do you have enough information to predict the price if the Smith family decides to get five phones? Explain.

7) Write a summary of your conclusions from this lesson.

---

[34] Retrieved May 13, 2013, from
http://www.uscellular.com/uscellular/plans/showPlans.jsp?type=plans&plan-selector-type=family.

Lesson 16, Part D, Where do we start?   Theme: Personal Finance

Let's see what else we can determine about the cell phone "family plans."

Number of phones	2	3	6
Cost per month ($)	99.99	109.99	139.99

Credit: Joshua Resnick/Fotolia

1) In Part C, you used the first two data points in the table to determine the slope of the relationship. Given that cost per phone, what is the total <u>phone</u> cost for the three phones?

2) How is the answer to question 1 related to the second data point in the table?

Objectives for the lesson

You will understand:

☐ The meaning of *y*-intercept in problem situations.

You will be able to:

☐ Determine the *y*-intercept in a linear relationship.

☐ Determine the equation of a linear relationship.

3) Create a coordinate plane like the one shown below, then graph all of the data points from the cost table. Draw a line that would include all of the points. Extend the line to the edges of the plane.

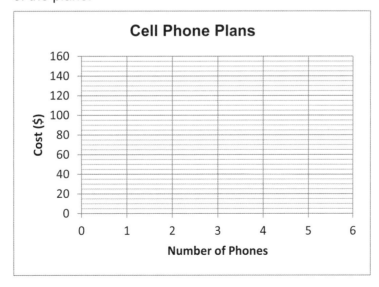

4)   Use the line to estimate the cost of having a plan with only one phone. How does this relate to question 2?

5)   Write a sentence that includes all of the information you need to know about the cost of cell service from this company.

6)   Now let's use shortcut language, replacing some words with mathematical symbols:
      Cost =

7)   Recall from the preview questions in your assignments that points on the coordinate plane are often called (x, y) and that the vertical axis is often called the y-axis. The point where a graph touches or crosses the y-axis is often called the **y-intercept**. Label the y-axis and the y-intercept on your graph. Does the value of the y-intercept make sense in this situation? Explain.

8)   The Smiths are also thinking about adding some premium channels to their cable package. They surveyed some friends and discovered the information shown in the table. Inspect the table to determine the slope of the relationship between the cost of cable service and the number of premium channels.

Number of premium channels	1	4	5
Cost	$54.98	$99.95	$114.94

9)   What is the cost per premium channel? What does this tell you about the y-intercept?

10)   Write a sentence in context, which includes the slope and y-intercept of this   relationship. Does the y-intercept make sense in this situation? Explain.

11)   Predict the cost of cable service if the Smiths order six premium channels. Explain.

Lesson 16, Part E, Predicting costs   Theme: Personal Finance

Companies who provide home services often charge for the service call in addition to an hourly labor charge. The table below shows the times and costs for an assortment of jobs completed by Spark's Electric.

Credit: Wendy Kaveney/Fotolia

1)   Write a sentence that represents the relationship between cost and hours for Spark's Electric services.

Objectives for the lesson

You will understand:

☐ The meaning of slope and *y*-intercept in problem situations.

You will be able to:

☐ Calculate the *y*-intercept of any linear relationship given two points.

☐ Write an equation that represents a linear relationship given two data points.

☐ Use the linear equation to make predictions.

The table lists the charges from a few of the jobs that Spark's completed last month.

Hours	2	5	11	16
Cost	$161	$335	$683	$973

2)   Use the first two points to determine the slope of Spark's relationship between cost and hours. Be careful with the order and location of the values! Express your answer as a unit rate with correct units labeled.

3)   Now use the first point to work with, and substitute those values and your slope into the mathematical sentence you wrote in question 1. Be sure to substitute the numbers into the correct locations. Simplify the righthand side as much as possible (you should still have one variable on that side).

4)   Now solve your simplified equation to determine the charge for service calls. What ordered pair (with units) does this charge represent?

5) Rewrite your mathematical sentence from question 1, this time replacing the hourly rate and the service call charge.

6) Spark's has a big job coming up that is estimated to take 32 hours to complete. Use your equation from question 5 to determine the estimated cost of the job.

7) Graph a line using the points from the table, then extend the line to check the value you found in question 4.

Hours	2	5	11	16
Cost	$161	$335	$683	$973

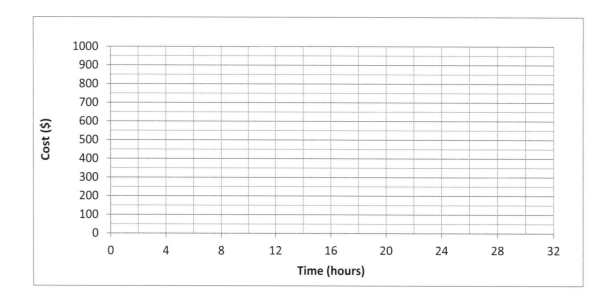

8) What advantage(s) does the graph of the line have over the equation or table representations of the line for solving problems?

9) What advantage(s) does the equation have over the graph?

Lesson 17, Part A, Expressing linear relationships

Theme: Civic Life

There are four useful ways to represent most mathematical relationships: graphs, tables, verbal descriptions, and equations. Each way has advantages and drawbacks. The graphs below model two approximately linear relationships.

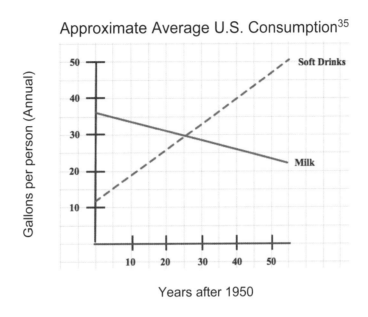

Approximate Average U.S. Consumption[35]

Years after 1950

1)   Write down three general conclusions you can draw from the graph.

Objectives for the lesson

You will understand that:

☐ Linear relationships can be expressed in different ways.

☐ There are advantages and disadvantages to each type of representation.

You will be able to:

☐ Express linear relationships in multiple forms.

☐ Identify appropriate use of each form of linear relationships.

☐ Compare two relationships.

2)   Now get more specific. Estimate at least 4 ordered pairs. Be sure to use each graph at least once.

---

[35] Adapted from www.ers.usda.gov/media/286050/sb965o_1_.pdf.

3)   Write contextual sentences about two of the points you estimated in question 2.

The table below is another representation of U.S. milk and soft drink consumption.

Approximate Annual Consumption

Years (since 1950)	Milk (gallons per person)	Soft Drinks (gallons per person)
0	36	11
15	32	22
30	28	33

4)   Calculate the slopes of the two lines. Write a sentence for each line describing the meaning of the slope.

5)   Now determine the equation for each line. Be sure to define your variables. Remember that your equation needs the slope and the *y*-intercept.

6)   During which time period did U.S. soft drink consumption surpass milk consumption? Explain how you drew this conclusion.

7)   If the soft drink linear pattern continued to hold true, when did/will consumption reach 50 gallons per person?

8)   If the milk linear pattern continued to hold true, when did/will milk consumption reach 20 gallons per person?

9)   Decide which representation (graph, table, verbal description, or equation) would be the most appropriate in each of the following situations.

Part A: You would like to display exact measurements for ten data points.

Part B: You are writing a magazine article about population and you would like to include a quick reference illustration of general trends.

Part C: You are talking to a friend on the phone and describing how two different companies calculate phone charges based on minute usage.

Part D: You would like to provide a short way for your friend from Part C to calculate his phone bill for any minute usage.

Summary Table

Representation	Strengths	Weaknesses
Graph		
Verbal		
Table		
Equation		

Lesson 17, Part B, Making the call   Theme: Personal Finance

Representations of linear relationships can be useful in decision-making. For example, you want to change cell phone providers and have to decide which plan would be best.

Plan 1 has a monthly fee of $15.99 plus $0.13 per minute.

Plan 2 costs $39.99 per month with unlimited minutes of talk.

Credit: Rido/Fotolia

1)   How many minutes would you have to talk to make Plan 2 worth the higher fee? Make a prediction.

Objectives for the lesson

You will understand:

☐ Linear relationships can be represented in various ways.

☐ Each representation of a line can be used to interpret the relationship.

You will be able to:

☐ Translate between representations of a line.

☐ Make decisions based on different forms of linear equations.

2)   Use the verbal descriptions above to create a table of costs depending on the number of minutes used.

Minutes	Plan 1 Cost	Plan 2 Cost
0		
50		
100		
150		
200		
250		
300		

3)   Write linear equations representing the cost of each plan.

4)   Use the equations you found in question 3 to find the number of minutes a customer would have to talk to have the same cost under both plans.

5)   Graph the lines that represent the costs of each plan.

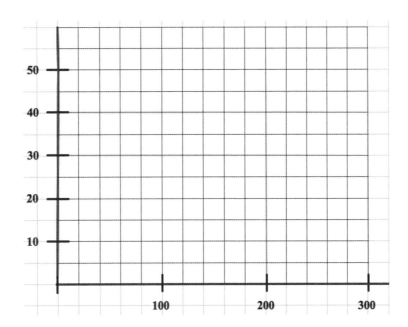

6)   Do your answers in questions 2 through 5 confirm or conflict with each other? Explain.

7)   Look back at question 1. How close was your prediction?

8)   Make an estimate of your own personal monthly phone use. Decide which of the phone plans would be better for you. Explain the reason for your choice.

9)   Pair with at least two other students and brainstorm at least two other types of monetary decisions you could make using the methods in this activity. Be prepared to share your ideas with the rest of the class.

| **Lesson 17, Part C, Close enough?** | **Theme: Civic Life** |

Throughout the twentieth century, life expectancies steadily increased.

Services such as Social Security and Medicare use long-term data about life expectancies to develop models that help them budget for longer term care.

The table to the right shows data for the life expectancy of a person born in the United States in years after 1900.[36]

1) Has life expectancy increased at a steady rate since 1900? How do you know?

Birth Year (after 1900)	Life Expectancy (years)
0	47.3
10	50.0
20	54.1
30	59.7
40	62.9
50	68.2
60	69.7
70	70.8
80	73.7
90	75.4
100	76.8

Objectives for the lesson

You will understand:

☐ Linear equations can be used to model approximately linear data.

☐ How to identify approximately linear data from a graph.

You will be able to:

☐ Identify approximately linear data.

☐ Write a linear equation to model approximately linear data.

2) Which decade appears to have the greatest gains in life expectancy? Which decade appears to have the least?

---

[36] U.S. Department of Health and Human Services. (2012, September 24). United States life tables. National Vital Statistics Reports, *61*(3). Retrieved June 17, 2014, from http://www.cdc.gov/nchs/data/nvsr/nvsr61/nvsr61_03.pdf.

3)   Pick any two data points and find the slope of the line containing them. Make sure to include units with your slope.

4)   Pick another pair of points and find the slope. If your two slopes are not the same, average them together. (Note: This is not always allowable—more discussion later.) Write a contextual sentence explaining your slope.

Even if a relationship is not perfectly linear, it might be appropriate to model the data with a linear equation. This equation represents a **trend line** that shows the general trend of the data. Three **scatterplots** are shown below. The trend line for the first graph is provided. Notice how the trend line is close to most points, and there are about the same number of points above and below.

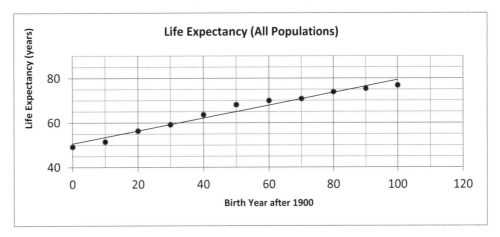

5)   Use a straight-edge to draw a line that best represents the trend in each graph shown below. The line should be close to all of the points, with about the same number of points above and below.

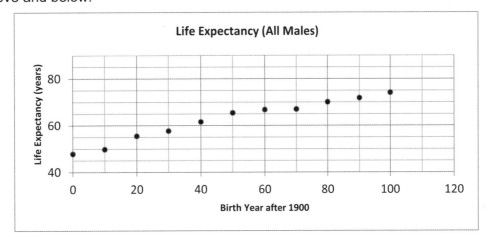

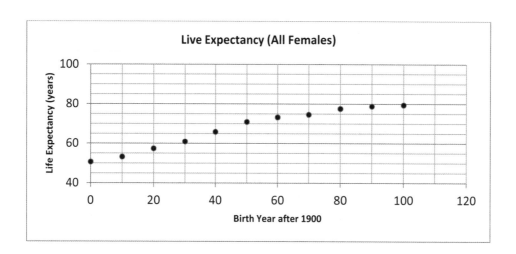

6) Use the trend line you drew on the Male Life Expectancy graph to estimate a slope and *y*-intercept. Use them to write an equation of the trend line.

7) Use the trend line you drew on the Female Life Expectancy graph to estimate a slope and *y*-intercept. Use them to write an equation of the trend line.

8) Write one or two sentences comparing the slopes of the trend lines for male and female life expectancy. Also, compare how well the lines model the data—are the points fairly close to the line?

9) The scatterplots below show many different sets of data. Decide which ones you think might be reasonable to model with a linear equation.

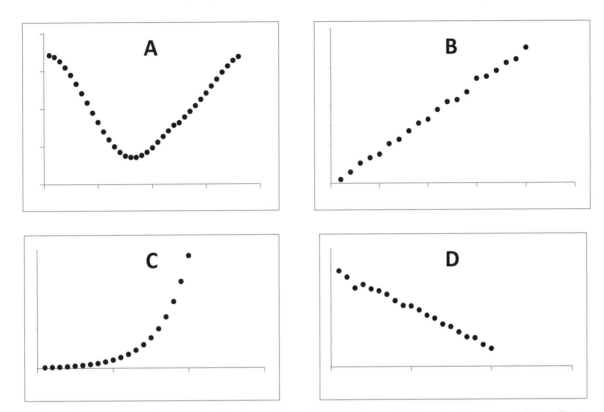

10) Consider the graphs you decided would not be good candidates for linear models. For each one, write down a description of the characteristics that helped you eliminate it. Be prepared to share your thoughts with the class.

Lesson 17, Part D, Predicting budget increases
Theme: Personal Finance

Home electricity usage is measured in **kilowatt-hours**.[37]

In this activity, you will apply what you have learned about linear relationships to analyze messy data and to use your analysis to make predictions about the financial impact of energy use decisions.

1) The table shows a sample of actual energy usage and monthly electric bill amounts for a small apartment at various times during the year 2012.[38] How does this data differ from the data in the previous activities?

Energy Use (kWh)	Bill ($)
198	38.34
699	114.84
366	63.43
741	117.35

Objectives for the lesson

You will understand that:

☐ Linear equations can be used to approximate relationships in some real data.

☐ Linear approximations of data can be used to make predictions beyond the data.

You will be able to:

☐ Write a linear equation to fit data.

☐ Use a linear approximation to predict expenses.

2) Create a graph of the data points, then use a straight-edge to draw a line that passes closely to all of the points.

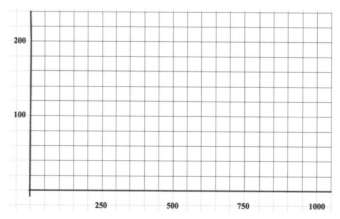

---

[37] A kilowatt-hour is a measure of electrical energy equivalent to a power consumption of one thousand watts for one hour; a watt is one joule per second (source: Oxford Dictionaries, www.oxforddictionaries.com).

[38] The amounts are from actual electric bills from a residence in Wichita Falls, Texas.

3)   Use the line you drew in question 2 to make an estimate of an electric bill in a month when the consumer used 500 kWh.

4)   Use the data for 198 kWh and 366 kWh to determine an equation for the line.

5)   Use the equation of the line you found in question 4 to estimate the bill when the usage is 500 kWh.

6)   Write down at least one sentence describing the advantages of each estimate (from questions 3 and 5).

7)   To the nearest cent, how much cost is added to the bill for every 1 kWh of electricity used?

8)   The tenant of the apartment decides to install a dishwasher. The model she chooses will use approximately 15 kWh every month.[39] Predict the change she can expect to see on her monthly electric bill.

9)   The tenant knows from past years that her electrical usage spikes in August due to the need for air conditioning during Texas summers. She estimates that her usage in August will be approximately 1,000 kWh. Using the linear equation you have created, estimate how much she should budget to pay her August electric bill.[40]

---

[39] Energy.gov. (n.d.) Estimating appliance and home electronic energy use. Retrieved June 17, 2014, from http://energy.gov/energysaver/articles/estimating-appliance-and-home-electronic-energy-use.

[40] Note that some electricity providers charge extra for high usage or during certain months.

Lesson 18, Part A, Pricing your product   Theme: Personal Finance

Credit: nenetus/Fotolia

In the Preview Assignment, you read about Eduardo and Paula. They are starting a small business making and selling tie-dyed clothing, and they disagree about the formula needed to add a 65% markup to the cost of the clothing.

Recall that B2 = the amount they pay to buy the V-neck tee (in dollars)

Eduardo thinks the formula should be =B2*1.65.

Paula thinks it should be =B2+0.65*B2.

1)   What did you decide about who was right?

Objectives for the lesson

You will understand that:

☐ Using algebra to generalize calculations can save time and promote accuracy.

You will be able to:

☐ Write and interpret an expression that models a percentage increase or decrease.

☐ Use expressions and equations to solve contextual problems about percentage increase or decrease.

☐ Solve problems with shifting reference values.

2)   Calculate the sales price (price after the markup has been included) for each item shown in the spreadsheet below.

	A	B	C
1	Item	Cost of Item	Sales Price
2	V-neck tee	$8.79	
3	Round neck tee	$10.13	
4	Child's t-shirt	$6.59	
5	Skirt	$23.45	
6	Dress	$35.67	
7	Scarf	$9.34	
8			

Six months later, Eduardo and Paula are doing very well with their business. They decided after the first event that prices to the nearest cent are not very practical because it takes so much time to make change for customers. Paula revised the spreadsheet to calculate the markup and then round the prices to the nearest dollar. The rounded results and the formula she used are shown below. (The number at the end of the formula indicates the number of decimal places desired.)

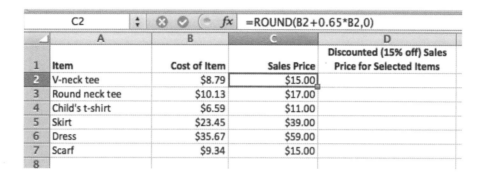

C2				fx	=ROUND(B2+0.65*B2,0)	

	A	B	C	D
	Item	Cost of Item	Sales Price	Discounted (15% off) Sales Price for Selected Items
1				
2	V-neck tee	$8.79	$15.00	
3	Round neck tee	$10.13	$17.00	
4	Child's t-shirt	$6.59	$11.00	
5	Skirt	$23.45	$39.00	
6	Dress	$35.67	$59.00	
7	Scarf	$9.34	$15.00	
8				

3) Although their products have sold very well, Paula and Eduardo have some items that have not sold. They decide to put these items on sale for 15% off.

Part A: Estimate the sale price of the items. Try not to use calculators or paper and pencil! See how many estimation strategies your group can describe.

Part B: Write a formula that Paula can use in Cell D2 to find the new sale price of a V-neck tee. Try to find more than one formula that works.

Part C: Paula and Eduardo have a few scarves that do not sell even after being discounted. They decide to take another 15% off. Eduardo says the result is the same thing as a 30% discount, but Paula disagrees. Who is right? What is the new sale price?

Lesson 18, Part B, Backing out the sales tax

Theme: Personal Finance

Credit: Monkey Business/Fotolia

Occasionally, an advertisement will say:

"We pay the sales tax for you!"

1) What is the difference between this and a sales tax holiday?

Objectives for the lesson

You will understand that:

☐ "Undoing" a percentage increase is not the same as a decrease by the same percentage.

You will be able to:

☐ Write and interpret an expression that models a percentage increase or decrease.

☐ Use expressions and equations to solve contextual problems about percentage increase or decrease.

2) At some events, Eduardo and Paula have to charge sales tax. This means that the total charge is no longer rounded to the nearest dollar, and they have to take the time to calculate the customers' change again. For example, for a V-neck tee, they need to charge about $15 to make a profit. With 7.5% tax, the total cost to the customer is $16.13. Paula decides to set the prices so that they come out to even dollar amounts after the tax is added. So, for the V-neck tee, they want the total after tax to be $17. (Eduardo and Paula decide to adjust the amount up to the next dollar so they do not lose any money.)

The spreadsheet below gives the amounts Eduardo and Paula want to charge after tax.

Part A: Calculate the price before tax that will result in these tax-included amounts. (The taxing agency requires business owners to report pre-tax prices, not the tax-included prices.) The tax rate is 7.5%.

	A	B	C
1	Item	Price before tax	Price after tax
2	V-neck tee		$17.00
3	Round neck tee		$19.00
4	Child's t-shirt		$12.00
5	Skirt		$42.00
6	Dress		$64.00
7	Scarf		$17.00
8			

Part B: What formula could Paula use in Cell B2 to calculate the pre-tax price for a V-neck tee?

3) The following questions give you more practice with using percentages in expressions and equations. Column A gives expressions that either increase or decrease the quantity x by a given percentage. Column B gives verbal descriptions of the changes. Write the letter of each description next to the correct expression. A verbal description may be used more than once or not at all.

Column A	Column B
1. $x + 0.45x$	a) Decrease x by $0.55
2. $0.45x$	b) Increase x by 45%
3. $x - 0.55x$	c) Increase x by 55%
4. $1.45x$	d) Increase x by 145%
5. $3.45x$	e) Increase x by $1.45
6. $2.45x$	f) Increase x by 245%
7. $x + 0.55x$	g) Increase x by 345%
8. $x - 0.45x$	h) Decrease x by 55%
9. $0.55x$	j) Decrease x by 45%
	k) Decrease x by 145%

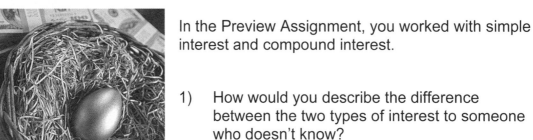

Credit: Andy Dean/Fotolia

In the Preview Assignment, you worked with simple interest and compound interest.

1) How would you describe the difference between the two types of interest to someone who doesn't know?

Objectives for the lesson

You will understand that:

☐ Compounding is a repeated multiplication by a compounding factor.

☐ Exponential growth models the compounding of interest on an initial investment.

You will be able to:

☐ Calculate the earnings on a principal investment with annual compound interest.

☐ Write a formula for annual compound interest.

2) Suppose you invest $1,000 **principal** into a certificate of deposit (CD) with a five-year **term** that pays a 2% annual percentage rate (APR) interest. The **compounding period** is one year.

Part A: How much will you have in the account at the end of the first year? The second year? The third year? Organize your work in the table below.

Term	Calculation	Amount Accrued
1 year		
2 years		
3 years		

Part B: Can you determine the amount in the account at the end of the fifth year without having calculated the fourth year? Explain.

3) Using patterns discovered in answering question 2, develop an equation for the total amount accrued in a CD with annual compounding after n years if the principal is $1,000 and the APR = 2%.

Lesson 18, Part D, Long-term growth   Theme: Personal Finance

Credit: Jerry Sliwowski/Fotolia

Consider your formula from Lesson 18, Part C.

1) How would you revise the formula if you had $50,000 to put in the CD and the bank was paying 2.5% APR (interest) compounded annually?

Objectives for the lesson

You will understand that:

☐ Compounding is a repeated multiplication by a compounding factor.

☐ Exponential growth models the compounding of interest on an initial investment.

You will be able to:

☐ Calculate the earnings on a principal investment with annual compound interest.

☐ Write a formula for annual compound interest.

☐ Compare and contrast linear and exponential models.

Generally, when your CD term is up, you can continue to leave the money in the account, but the bank may change the APR at that time. Let's assume the bank doesn't change the rate.

2) Use the formula from question 1 to determine how much you would have in the account in question 1 at the end of 20, 50, and 100 years.

Time	Calculation	Amount Accrued
20 years		
50 years		
100 years		

3)   Is your formula from question 2 linear? Explain your reasoning.

4)   Write a general formula that could be used to find the accrued amount (A) for a CD with annual compounding interest. Let P = the principal, r = the APR as a decimal, and n = the number of years.

Remember that, in actuality, CD rates will normally adjust at the end of the term, so forecasting 20, 50, or 100 years into the future is unrealistic. However, investors can purchase municipal bonds that last for years. The *Nashua Telegraph*[41] (New Hampshire) reported a case in 1981 in which a man found a $100 bond that had been accruing interest at 10% for 110 years!

_____

[41] Old county bond worth millions. (1981, June 27). *Nashua Telegraph*, p. 12. Retrieved June 15, 2014, from
http://news.google.com/newspapers?nid=2209&dat=19810627&id=AKsrAAAAIBAJ&sjid=PPOFAAAA
IBAJ &pg=6938,6045283.

Lesson 19, Part A, More compounding   Theme: Personal Finance

Suppose you invest $1,000 principal in a two-year CD with an annual percentage rate[42] of 6%, where compounding occurs monthly.

1)   If the annual rate is 6%, what is the monthly interest rate?

Credit: Michalis Palis/Fotolia

2)   Over the two-year term of the CD, how many times will the bank pay interest to you?

Objectives for the lesson

You will understand:

☐ The characteristics of exponential growth.

You will be able to:

☐ Use the compound interest formula for different compounding periods.

3)   Use your answers to questions 1 and 2 to complete the following table. Use the middle column to record how you found the results.

- Principal = $1,000
- Interest = 6%
- Term = 2 years
- Compounding period = Monthly

Period	Calculation	Amount Accrued
1 month		
2 months		
3 months		
6 months		

---

[42] Financial institutions usually advertise the APY (annual percentage yield) rather than APR. The APY is the effective interest the investor receives and includes the effect of the compounding.

4)   Write a general formula that can be used to calculate the value of any CD. Define your variables.

5)   If compounding is done more often, what happens to the accumulated amount? Be specific.

Lesson 19, Part B, Depreciation   Theme: Personal Finance

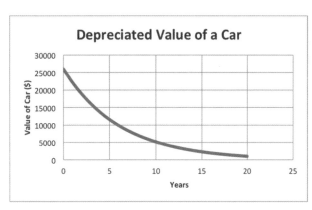

Depreciated Value of a Car

Depreciation is a process of losing value.

For example, new automobiles typically lose between 15% and 20% of their value each year for the first few years after manufacture.

1)   Compare depreciation of a car with interest accrued on a certificate of deposit.

Objectives for the lesson

You will understand:

☐ The differences and similarities between exponential growth and exponential decay.

You will be able to:

☐ Write an exponential decay model.

2)   How much is a $26,000 car worth after 1 year if the depreciation is 15% per year?

3)   Develop a value formula ($V =$) for a $26,000 automobile based on 15% depreciation per year. The table below can be used as a guide to your calculations.

Age of the Automobile	Calculation	Value
New		
1 year old		
2 years old		
5 years old		
t years old		

4)   Based on your formula, and assuming that depreciation continues at the same rate, what will the car be worth in 10 years? After finding the value, check for reasonableness by comparing to the graph.

Lesson 19, Part C, Payday loans   Theme: Personal Finance

In the Preview Assignment, you read about **payday loan stores.**

The average payday loan charges 13% interest for a two-week period.

1) What is the repayment amount for a $100 payday loan at these terms? Explain.

Credit: creative soul/Fotolia

Objectives for the lesson

You will understand that:

☐ Compound interest can be applied to short-term loan scenarios.

You will be able to:

☐ Build an exponential model.

2) Sometimes borrowers find themselves unable to pay back the original loan so they "rollover" the loan and interest into a new two-week loan. What is the repayment amount for the new loan?

3) If this continued for a year, how many two-week periods will have passed? What would be the total amount owed?

4) Create a table or graph that shows the relationship between the repayment amount and the number of two-week periods that have passed. Include at least 6 data points.

Lesson 19, Part D, Neither a borrower . . .
Theme: Personal Finance

Shakespeare said, "Neither a borrower nor a lender be; for loan oft loses both itself and friend."[43]

1)   What was Shakespeare trying to say?

Credit: Alexandra Thompson/Fotolia

Objectives for the lesson

You will understand that:

☐ Linear and exponential growth differs significantly in the long run.

You will be able to:

☐ Compare an exponential and linear model, and make a decision based on the comparison.

In the previous activity, you investigated payday loan centers. Another option in an emergency is to ask a family member or a friend. Suppose a friend offers to lend you $100 if you agree to repay the original amount, plus $10 for each week you owe her any money, until all the money is paid back.

2)   If you pay the loan off completely at the end of the week, how much is the payment?

3)   If you wait and pay her at the end of one year, how much will the payment be?

4)   Create a table or graph that shows the relationship between the repayment amount and the number of weeks that have passed. Include at least 6 data points.

5)   If you pay her $10 per week on the principal and also pay that week's interest, how long will it take to pay back the entire loan? How much will you have paid altogether?

---

[43] Shakespeare, W. (n.d.) *The Tragedy of Hamlet, Prince of Denmark.* Act I, Scene 3. Retrieved June 15, 2014, from http://shakespeare.mit.edu/hamlet/hamlet.1.3.html.

Lesson 19, Part E, Credit card repayment

Theme: Personal Finance

Finance charges on credit cards are calculated by several different methods depending on the card and the company. None of these methods are exactly exponential, but like payday loans, an exponential model can be used to approximate the charges.

For example, finance charges based on a credit card rate of 24.99% is about equivalent to a 28.06% annual rate compounded daily.

Credit: sumire8/Fotolia

1) Calculate the interest for one year on $2,000 with an APR of 28.06% compounded daily. Note you are asked for the interest daily, not the total balance.

Objectives for the lesson

You will understand that:

☐ Credit card interest can be approximated with an exponential model.

You will be able to:

☐ Compute and compare repayment schedules.

2) Investopedia illustrates credit card interest with the following example:

John and Jane both have $2,000 of debt on their credit cards. Each month, John and Jane are charged a 20% annual interest on their cards' outstanding balances. So, when John and Jane make payments, part of the payments goes to paying interest and part goes to the principal.[44]

Part A: Assume the interest is compounded monthly. Calculate the interest charged for 1 month for John and Jane (it is the same for both). Round your answer to the nearest cent.

---

[44] Understanding credit card interest. In *Investopedia.* Retrieved June 15, 2014, from www.investopedia.com/articles/01/061301.asp#axzz1bo1CMcfb.

Part B: The cards each require a minimum monthly payment of 3% of the balance owed, or $10, whichever is higher. Both are strapped for cash, but Jane manages to pay an extra $10 on top of her minimum monthly payments. John pays only the minimum. What is the minimum payment for the first month? How much of this amount will go to reduce the balance?

Part C: John pays the minimum and Jane pays an extra $10. Part of this pays the interest you calculated in Part A, and the rest pays off the balance. Complete the table below. Round to the nearest cent.

	John	Jane
Original balance	$2,000	$2,000
Amount paid on the balance		
New balance after the payment is subtracted		

Part D: Calculate the interest that John and Jane each pay in the second month. Remember to use the new balance for each. Round to the nearest cent.

John's interest:

Jane's interest:

You might wonder if the small difference in the interest charges for John and Jane really matters. Investopedia gives the following summary if John continues to make the minimum payment and Jane continues to pay an extra $10 each month: It takes John 15 years to pay off the original debt of $2,000, and he pays a total of $4,240. Jane pays off her balance in 7 years and pays a total of $3,276.

Part E: How much interest did John pay? What percentage of John's total payment is for interest?

Part F: How much interest did Jane pay? What percentage of Jane's total payment is for interest?

If you are interested in reading the full Investopedia article, you can find a link in the footnote on the first page of these student pages.

If you are interested in how long it will take you to pay off a credit card if you only make the minimum payment, visit the U.S. Federal Reserve website at: http://www.federalreserve.gov.

Resource **Overview**

Each assignment in the course will be structured around the principles of:

- Developing skills and understanding.
- Making connections to prior learning.
- Preparing for future lessons.
- Reading, writing, and reflection.

Each assignment will contain problems in which you practice the skills from the lesson, including extending those skills in a new way or applying them to a new situation.

Taking control of your own learning

In addition, each assignment will include questions designed to prepare you for future lessons. You will be given resources to help you, if needed.

When you use videos as a tool for refreshing on concepts, it is best to actively engage with the material in the video, rather than watching passively. For example:

- Copy the examples.

- Pause the video and work problems when directed to do so.

- When you have to correct your work, do so underneath the original work and write an explanation of your error and/or the correction.

- Watch the video a second time and add comments by the examples.

Self-regulating your learning

One goal of this course is to increase your ability to learn efficiently and effectively. This means learning faster and learning smarter—what scientists call being a "self-regulated learner." The following section explains what this means.

Self-regulating your learning means you *plan* your work, monitor your *work* and progress, and then *reflect* on your planning and strategies and what you could do to be more effective. These are the three phases of Self-Regulated Learning (SRL). They are introduced as follows and will be followed up on later in the course.

> **Plan:** Before doing a problem or assignment, self-regulated learners *plan*. They think about what they already know or do not know, decide what strategies to use to finish the problem, and plan how much time it will take. Research has shown that math experts often spend much more time planning how they will do a problem than they do actually completing it. Novices, the people who are just starting out, often do the opposite.

Work: Self-regulated learners use effective strategies as they **work** to solve problems. They actively *monitor* what study strategies are working and make changes when they are not working. When they do not know which strategy would be better, they ask for help. Self-regulated learners also keep themselves focused while they are working and pay attention to their feelings to avoid getting frustrated.

Reflect: Usually after an assignment or problem is done, self-regulated learners take time to reflect about what worked well and what did not. Based on that reflection, they think about how to change their approach in their future. The ***reflect*** phase helps self-regulated learners understand more about how they learn so they can become more efficient and more effective the next time. Reflecting is important for doing a better job next time you plan for a new problem or assignment.

You can think of these three phases as a cycle. You incorporate what you learned during the reflect stage in your *next* plan phase, making you a more effective learner as you repeat this process many times. The most effective students get in the habit of working this way:

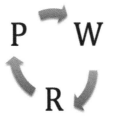

For most people, self-regulating takes time, practice, and hard work, but it is always possible. People can improve even if, in the beginning, they did not self-regulate their learning very well. The more you practice something and the more you train your brain to think in certain ways, the easier it becomes.

Resource **5-Number Summary and Boxplots**

5-number summary

In your previous course, you learned about the **5-number summary** for describing a set of data. Consider the following set:

550, 61, 75, 228, 79, 121, 79, 129, 240, 150, 147, 72, 142, 50

Step 1: Put the values in order from smallest to largest and identify the smallest value (the minimum) and the largest value (the maximum).

50, 61, 72, 75, 79, 79, 121, 129, 142, 147, 150, 228, 240, 550

The minimum value is 50 and the maximum value is 550.

Step 2: Find the median.

50, 61, 72, 75, 79, 79, 121,   129, 142, 147, 150, 228, 240, 550

↑

The median is (121 + 129)/2 = 125

Step 3: Find Q1 and Q3.

Find the median of the smaller half of the numbers:

50, 61, 72, 75, 79, 79, 121

Q1 is 75 ↑

Find the median of the larger half of the numbers:

129, 142, 147, 150, 228, 240, 550

Q3 is 150 ↑

Step 4: Report your 5-number summary in order.

Minimum	50
Q1 (first quartile)	75
Median	125
Q3 (third quartile)	150
Maximum	550

Box-and-whisker plots: Graphical displays of the 5-number summary

Box-and-whisker plots (often called simply "boxplots") are a graphical way of illustrating data in quartiles (groups of 25% of the data).

A basic box-and-whisker plot can be created using the steps described below.

Step 1: Create a number line with evenly spaced increments, which reaches below the minimum value and above the maximum value of the data set.

| 50 | 100 | 150 | 200 | 250 | 300 | 350 | 400 | 450 | 500 | 550 |

Step 2: Place vertical markers above each of the values in the 5-number summary from the previous page.

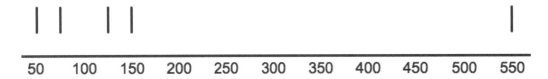

| 50 | 100 | 150 | 200 | 250 | 300 | 350 | 400 | 450 | 500 | 550 |

Step 3: Draw horizontal markers to create the box and the whiskers as shown.

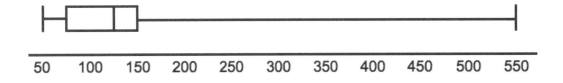

| 50 | 100 | 150 | 200 | 250 | 300 | 350 | 400 | 450 | 500 | 550 |

Step 4: Analyze the result.

Sample observation: Notice that the whisker on the right side is very long; this means that the top 25% of the data values are very spread out. The left whisker is very short; the lowest 25% of the data values are very close together.

Remember, this is the basic boxplot format. More sophisticated boxplots can be used to highlight unusual values in the data.

Resource **Algebraic Terminology**

Some notes about algebraic terminology

Formulas are a type of an **algebraic equation**. You have probably seen algebraic equations in previous math classes. For example:

$$y = x + 3$$

Each side of this equation is called an **algebra expression**.

$$\overbrace{y}^{expression} = \overbrace{x+3}^{expression}$$

So "$x + 3$" is an expression and "y" is an expression. The equal sign indicates that the two expressions are equal, thus forming an **equation**.

Therefore, an equation must have an equal sign with expressions on each side.

Expression = Expression

[Note that this is exactly the same as $x + 3 = y$]

An equation defines a **sequence of calculations**, often using algebra to shorten the information. In the example above, this sequence is:

1) Start with x

2) Add three to x

3) The result is y

Notice how much shorter $y = x + 3$ is than the three listed steps.

The word **formula** is usually used to express important and non-changing relationships, especially in contexts such as science, business, medicine, sports, or statistics. For example, in Lesson 11, you used the formula for the area of a rectangle, $A = L \cdot W$. This is a formula because the relationship between area and the length and width of a rectangle is always the same.

An example of an equation would be if you had a job in which you make $12 per hour. This relationship could be written algebraically as $P = 12h$ where P is your pay in dollars, and h is the number of hours you work. If you get a raise, the relationship would change. You also might call the equation a **model** because it models a situation using mathematics.

Resource **Coordinate Plane**

Vocabulary

Axes: A coordinate plane has two axes that measure distance in two dimensions. The *horizontal axis* goes from left to right. In previous classes, you may have called this the *x-axis*. The *vertical axis* goes up and down. This is sometimes called the *y-axis*. The axes are two number lines that create a grid on the coordinate plane. **Note:** *Axis* is singular and *axes* is plural.

Origin: The point at which the two axes intersect or cross is called the origin. This point represents 0 for both axes. To the left of this point, the horizontal axis is negative; to the right, it is positive. Below the origin, the vertical axis is negative; above the origin, it is positive. You can see this in the numbers along each axis. These numbers are called the *scale*.

Ordered pair: Each location or point on the coordinate plane is defined by an ordered pair. You can think of this as the "address" of a point. Ordered pairs are written in a set of parentheses. They are called *ordered pairs* because they must contain two numbers and the order of the numbers is important. The first number is the distance and direction going left or right from the origin, and the second number is the distance and direction going up or down. The ordered pair for the origin is (0, 0).

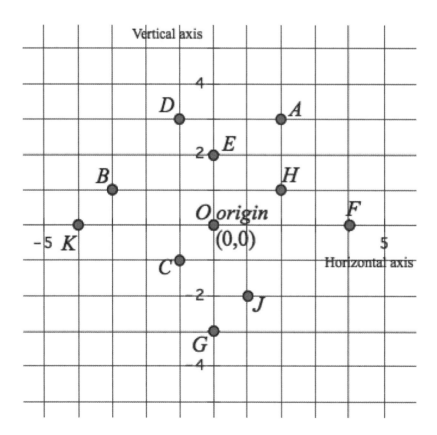

Follow these steps to find the point represented by the ordered pair (2, 3):

Step 1   First, think about the address of the point. If this were a street address, the ordered pair tells you to walk 2 blocks horizontally in the positive direction (right) and then walk 3 units vertically in the positive direction (up).

Step 2   Start at the origin. Go 2 units to the right because this is the positive side of the horizontal axis.

Step 3   Go 3 units up.

Point A on the graph is the point (2, 3). A few other examples from the graph are given below:

Point B: (−3, 1)   Point E: (0, 2)   Point F: (4, 0)

Where would you place point P (−4, 5)?

Resource **Dimensional Analysis**

Quantitative reasoning skill: Ratios and unit rates

Unit rates are ratios with a denominator of 1, although they are not always written as fractions.

For example, 60 mph is the same as $\dfrac{60 \text{ miles}}{1 \text{ hour}}$.

The language "miles per hour" implies that the operation is miles divided by 1 hour.

As another example, in 2012, the federal minimum wage was $7.25/hour. This means that an

employee earns $7.25 for 1 hour of work, or $\dfrac{\$7.25}{1 \text{ hour}}$.

A worker may also be paid a weekly salary. If an advertisement states that a job pays $320 for a 40-hour work week, then that position can be compared to the previous job by converting to a unit rate:

$$\dfrac{\$320}{40 \text{ hours}} = \dfrac{8 \cdot 40}{1 \cdot 40} = \dfrac{\$8}{1 \text{ hour}} \qquad \text{The second job pays better.}$$

Another way to think about the calculation above is as a division problem: $320 \div 40 = 8$

This can be helpful when the numerator and denominator do not have a common factor.

Quantitative reasoning skill: Conversion factors

In the above example, the fraction was simplified by dividing out the common factor of 40/40 (which is equivalent to 1). A fraction that is a ratio of quantities can be equivalent to 1 even when the numerator and denominator are not the same number. However, it is necessary that the numerator and denominator represent equivalent quantities. For example, the following fractions are all forms of one:

$$\dfrac{16 \text{ ounces}}{1 \text{ pound}} \qquad\qquad \dfrac{1 \text{ mile}}{5{,}280 \text{ feet}} \qquad\qquad \dfrac{60 \text{ minutes}}{1 \text{ hour}}$$

These types of ratios are sometimes called conversion factors because they can be used to covert between units.

The example below shows how to set up a multiplication problem with the rate and the conversion factor to convert miles per hour to miles per minute.

$$\dfrac{35 \text{ miles}}{1 \,\boxed{\text{hour}}} \cdot \dfrac{1 \,\boxed{\text{hour}}}{60 \text{ minutes}}$$

Notice that the conversion factor is written so that the units of hours are in the numerator. This is because you want the *hours* label to divide out in the same way that common factors divided out in the weekly salary problem above. This leaves the units of miles/minute as shown here:

$$\rightarrow \dfrac{35 \text{ miles}}{1 \,\cancel{\text{hour}}} \cdot \dfrac{1 \,\cancel{\text{hour}}}{60 \text{ minutes}} \quad \rightarrow \quad \dfrac{35 \text{ miles}}{60 \text{ minutes}} \quad \rightarrow \quad \dfrac{0.58 \text{ mile}}{1 \text{ minute}}$$

Before continuing, make sure you can answer the following question:

- How was the 0.58 calculated?

Quantitative reasoning skill: Dimensional analysis

Dimensional analysis, unit analysis, or unit conversion are all names for the process of using conversion factors to set up and solve certain types of problems. Many professionals—including pharmacists, dietitians, lab technicians, and nurses—use unit analysis. It is also useful for everyday conversions in cooking, finances, and currency exchanges. Many people can do simple conversions without dimensional analysis; however, they will likely make mistakes on more complex problems.

The advantage of using dimensional analysis is that it is a way to check your calculations. While it is always important that you develop your own methods to solve problems, this is a time when you are encouraged to learn and use a specific method. Once you have learned dimensional analysis, you can decide when to use it and when to use other methods.

In Lesson 1, Part C, you needed to convert inches into feet and then into miles. The computation below shows how dimensional analysis can help you organize your work.

Example 1:

$$\frac{1,000 \text{ people}}{1,500 \text{ feet}} \times \frac{5,280 \text{ feet}}{1 \text{ mile}} =$$

Unit labels can "cancel" in the same way that common factors do.

$$\frac{1,000 \text{ people}}{1,500 \text{ feet}} \times \frac{5,280 \text{ feet}}{1 \text{ mile}} =$$

$$\frac{5,280,000 \text{ people}}{1,500 \text{ mile}} = 3,520 \text{ people per mile}$$

Multiply numerators.

Multiply denominators.

Simplify.

Remember that you must set up your conversion factor (the multiplier) so that matching labels appear in one numerator and one denominator.

Some more extensive examples are shown on the next page.

Example 2:

Here is an example converting 1 year into seconds:

$$\frac{365 \text{ days}}{1 \text{ year}} \times \frac{24 \text{ hours}}{1 \text{ day}} \times \frac{60 \text{ minutes}}{1 \text{ hour}} \times \frac{60 \text{ seconds}}{1 \text{ minute}} =$$

$$\frac{365}{1 \text{ year}} \times 24 \times 60 \times 60 \text{ seconds} = 31{,}536{,}000 \text{ seconds per year}$$

Example 3:

A nurse or a pharmacist might need to know how many tablespoons are in a 250-mL (milliliter) bottle of medication. In this case, we are converting metric units (milliliters) to U.S. units (tablespoons).

There are 250 mL in one bottle, and a quick Internet search indicates 1 tablespoon is about 14.79 milliliters.

$$\frac{250 \text{ mL}}{1 \text{ bottle}} \times \frac{1 \text{ tablespoon}}{14.79 \text{ mL}} =$$

$$\frac{250 \text{ mL}}{1 \text{ bottle}} \times \frac{1 \text{ tablespoon}}{14.79 \text{ mL}} =$$

$$\frac{250}{14.79} \approx 16.9 \text{ tablespoons per bottle}$$

Example 4:

A father found some instructions on the Internet for building a treehouse. The instructions were in metric units but he only had a standard English ruler. The instructions said the boards for the framing should be 2.5 meters long. How many inches should the dad measure?

$$\frac{2.5 \text{ meters}}{1 \text{ board}} \times \frac{100 \text{ cm}}{1 \text{ meter}} \times \frac{1 \text{ inch}}{2.54 \text{ cm}} =$$

$$\frac{2.5 \text{ meters}}{1 \text{ board}} \times \frac{100 \text{ cm}}{1 \text{ meter}} \times \frac{1 \text{ inch}}{2.54 \text{ cm}} =$$

$$\frac{250}{2.54} \approx 98.4 \text{ inches per board}$$

Of course, if you can connect to the internet, you could just search for "free online conversion calculator"!

Resource **Equivalent Fractions**

Quantitative reasoning skill: Equivalent fractions

Two fractions are equivalent if they have the same value or represent the same part of an object.

For example, the figure shows that 1/2, 2/4, and 4/8 all represent the same part of a whole. They are equivalent fractions.

Recall that the *denominator* of a fraction represents the number of parts into which the whole has been divided. The *numerator* represents a count of the number of parts.

So, $\frac{4}{8}$ means that the whole is divided into 8 equal parts, and 4 of these parts are counted.

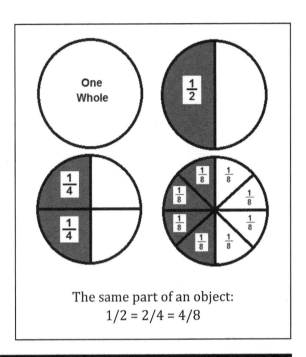

The same part of an object:
1/2 = 2/4 = 4/8

Quantitative reasoning skill: Simplifying fractions

The fraction 50/100 is equivalent to 1/2. Note that you can write:

$$\frac{50}{100} = \frac{1 \cdot 50}{2 \cdot 50} = \frac{1}{2} \cdot 1$$

The above calculation shows that both 50 and 100 can be written as a number times 50. You say that 50 and 100 have a "common factor of 50."

Another way to think of this is that the number 1 (written as $\frac{50}{50}$) is embedded in the fraction $\frac{50}{100}$.

"1" is a special number in mathematics because if you multiply any number by 1, you get a result that is equivalent to the original.

By dividing $\frac{50}{50}$ to get 1, you simplify $\frac{50}{100}$ to $\frac{1}{2}$.

In this case, the word *simplify* means that the fraction has been written in an equivalent form with smaller numbers. The *simplest* form means that the fraction is written using the smallest possible numbers. In general, answers should always be given in simplest form unless the question specifically calls for a different form.

Caution: It is common to say that you are writing the fraction in "reduced form." This language is misleading—the value of the simpler fraction is the same as the original fraction, but the word *reduced* implies that the "reduced fraction" represents a smaller quantity. The terminology *simplest form* or *lowest terms* makes more sense.

Resource **Four Representations of Relationships**

Mathematical relationships can be represented in four ways: models (equations), tables, graphs, and verbal descriptions. If you are struggling with a problem, approaching it with a different representation may help you to make sense of the work.

Model or equation

In Lesson 12, your class wrote a mathematical equation for the relationship of the price of gas and the cost of driving Jenna's car. An equation is useful because it can be used to calculate the cost values. As you saw with the formula for braking distance in Lesson 13, equations are also useful for communicating complex relationships. In writing equations, it is always important to define what the variables represent, including units. For example, in Lesson 12, the variables were defined as shown below. Note that each definition includes what the variable represents, such as *cost of Jenna's car*, and the units in which this quantity is measured, such as *$/mile*.

J = Cost of Jenna's car in $/mile

g = Price of gas ($/gal)

These variables were used in the mathematical equation, $J = \dfrac{g}{22} + 0.146$.

Table

Another way that you could have represented this relationship between the price of gas and the cost of driving the car is in a table that shows values of g and J as *ordered pairs*. An ordered pair is two values that are matched together in a given relationship. You used this representation in Lesson 13 when you explored how one variable affected another. Tables are helpful for recognizing patterns and general relationships or for giving information about specific values. A table should always have labels for each column. The labels should include units when appropriate.

Price of Gas ($/gal)	Cost of Driving Jenna's Car ($/mile)
3.00	0.28
3.50	0.31
4.00	0.33
4.50	0.35

Graph

In the last few assignments of Lesson 10, Part A through Lesson 14, Part D, you practiced with graphs. A graph provides a visual representation of the situation. It helps you to see how the variables are related to each other and make predictions about future values or values in between those in your table. The horizontal and vertical axis of the graph should be labeled, including units.

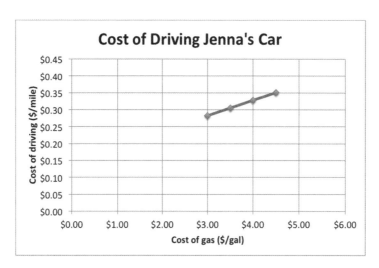

Verbal description

A verbal description explains the relationship in words, which can help you make sense of what the relationship means in the context.

For $J = \dfrac{g}{22} + 0.146$, the fraction $\dfrac{g}{22}$ represents the per-mile cost of the gas, which depends on

the price of the gas $\dfrac{\$}{1\ \text{gallon}} \cdot \dfrac{1\ \text{gallon}}{22\ \text{miles}} = \dfrac{\$}{\text{mile}}$ and 0.146 represents the per-mile costs

associated with oil changes, tire wear, etc. So the equation $J = \dfrac{g}{22} + 0.146$ represents the total

per-mile cost of Jenna's car. Verbally, Jenna might say, "I need to find the per-mile cost of my car so that I can compare it to the cost of a car rental. Dividing the gas price by 22 miles per gallon will give me the per-mile cost of the gas and then I need to add in the scaled costs of maintenance to get the total."

Summary

Throughout this course, you have learned that having the skill to move between different forms and tools is important in problem solving. Alternating among the four representations of mathematical relationships is another example of this. In some cases, you may struggle writing an equation, but find that starting with a table is helpful. You might want a graph for a visual representation, but also need to express a relationship in words. It is important that you can translate one form into another and also that you can choose which form is most useful in a specific situation.

Resource **Fractions, Decimals, Percentages**

Language of fractions, decimals, and percentages

There are several important vocabulary words you should know and use.

- A ratio is a comparison of two numbers by division. You will see many different types of ratios in this course. Some ratios are a special type called a percentage. A percentage is a ratio because it is a number compared to 100.
- Percentages are a relationship between two values: the *comparison value* and the *base value*. The relationship is described as a *percentage rate,* which is shown with a percentage symbol (%). This indicates that the rate is out of 100.

  Example: 10 is 20% of 50.

  10 is the comparison value.

  50 is the base value.

  20% is the percentage rate; it can be written as a decimal by using the relationship to 100: $\frac{20}{100} = 0.2$

- Fractions have two parts: $\frac{\text{numerator}}{\text{denominator}}$
- Every fraction can be written in *equivalent* forms (e.g., $\frac{2}{3} = \frac{4}{6} = \frac{6}{9}$). It is often useful to write the fraction in the form with the smallest numbers. This is called *simplified* or *reduced*. In the example, $\frac{2}{3}$ is in simplest form.

On the next page are some **common percentage benchmarks**.

Simplified Fraction	Percent	Decimal
$\frac{1}{100}$	1%	0.01
$\frac{1}{10}$		
		0.2
	25%	
$\frac{1}{3}$	round to the nearest one percent	round to nearest hundredth
		0.5
$\frac{2}{3}$	round to the nearest one percent	round to nearest hundredth
		0.75

Calculating percentage rates

Additional examples of finding the percent of a number and calculating percentage rate can be found at http://www.purplemath.com.

Resource **Length, Area, and Volume**

Length

Length is one-dimensional. An example would be the length of an extension cord that you need to plug in an electronic device. Examples of units of measure for length are inches, feet, yards, or miles (or in the metric system, centimeters, meters, or kilometers).

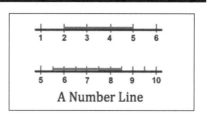
A Number Line

A number line can be used to model lengths.

The thicker segment on each number line shown above is 3 units long. If the scale is in inches,
each line segment is 3 inches long. If the scale is in feet, each line segment is 3 feet long.

Area

Area is two-dimensional and is measured in square units. The total number of one-foot square tiles needed to cover the floor of a room is an example of area measured in square feet, and can be modeled with a rectangle. Recall the formula for the area of a rectangle:

$$A = L \times W$$

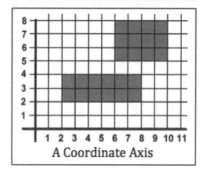
A Coordinate Axis

The area of a rectangle is the product of the length and the width, which is a shortcut for counting the number of square units needed to cover the rectangle.

Each of the two shaded areas on the coordinate axis has an area of 12 square units. If the horizontal and vertical scales are in inches, each area is 12 square inches. If the scales are in feet, each area is 12 square feet. Notice that the regions measured do not have to be squares, yet the area is measured in square units.

If the units are in inches, the area of the top rectangle is:

$A = (3 \text{ inches}) \times (4 \text{ inches})$

$= (3 \times 4) \times (\text{inches} \times \text{inches})$

$= 12 \text{ inches} \times \text{inches}$

$= 12 \text{ square inches}$

If the units are in feet, the area of the bottom rectangle is:

$A = (2 \text{ feet}) \times (6 \text{ feet})$

$= (2 \times 6) \times (\text{feet} \times \text{feet})$

$= 12 \text{ feet} \times \text{feet}$

$= 12 \text{ square feet}$

For more about labeling units, see the note on the next page.

Note 1:

It is common to abbreviate the units of measure using exponents. If the area is $A = 12 \ feet \times feet = 12 \ square \ feet$, we often write $A = 12 \ ft^2$

Notice the connection to algebra here!

Multiplying $(3 \, feet)$ by $(4 \, feet)$ is similar to multiplying $(3x) \times (4x)$. You multiply the numbers in front of the variables (coefficients), and then multiply the variables:

$$(3x) \times (4x) =$$
$$(3 \cdot 4)(x \cdot x) =$$
$$12x^2$$

Note 2:

It is common to confuse length and area formulas. Look at the bottom rectangle. Because it is shaded, it is tempting to think about area. If you want to know how many floor tiles to buy, area is the correct concept.

But what if you want to trim the edges of the room with baseboards? (If you aren't sure about this term, search the internet for "baseboard images.")

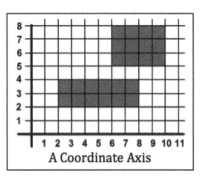

A Coordinate Axis

The length of the distance around a shape is called the **perimeter**. To calculate this length, simply add the total number of units as if traveling around the edge. For example, if the units are in feet, then the length of the line around the bottom rectangle is

$$P = 2 \, feet + 6 \, feet + 2 \, feet + 6 \, feet$$
$$= 16 \, feet$$

The arithmetic operation for length is **addition**, and the unit of measure is feet (you are adding up a lot of feet, so the final result is feet). By comparison, the arithmetic operation to compute area is **multiplication**, and the unit of measure is square feet (you are determining the number of square tiles). Again, this connects to algebra. To add algebraic terms, you must have **like terms**, meaning terms with the same variables:

$$2x + 6x + 2x + 6x =$$
$$16x$$

You cannot add $2x + 3y$ just as you cannot add $2 \, feet + 3 \, inches$.

Not every shape you need to find the length or the area of will be a square or rectangle. Think about a circular rug in the living room, or a gazebo in the shape of an octagon.

Circle length and area

The distance around a circle is called the **circumference**, which can be found with the formula:

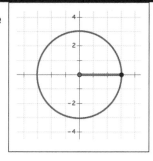

$$C = 2\pi r$$

The area of a circle is given by the formula:

$$A = \pi r^2$$

- π is a *constant* that is approximately 3.14159
  (You probably learned 3.14, but carrying additional decimal places reduces the amount of rounding error.)

- r is the radius of the circle, which *varies* depending on the size of the circle. It is the distance from the center to any point on the circle.

- C is the circumference of the circle, which *varies* depending on the radius.

- A is the area of the circle, which *varies* depending on the radius.

In this example, the radius is 3 units. Let's say those units represent inches.

The circumference is:

$$C = 2\pi r$$
$$C = 2\pi(3\ inches)$$
$$= 2{\cdot}3{\cdot}\pi\ inches$$
$$= 6\pi\ inches\ (exactly)$$
$$\approx 6{\cdot}3.14159\ inches$$
$$\approx 18.85\ inches$$

Look at the length of 1 unit on the radius. Does $19\ inches$ seem like a reasonable estimate for the distance around the circle?

The area is:

$$A = \pi r^2$$
$$A = \pi(3\ inches)^2$$
$$= \pi(3\ inches) \times (3\ inches)$$
$$= \pi{\cdot}3{\cdot}3(inches)(inches)$$
$$= 9\pi\ square\ inches$$
$$\approx 9{\cdot}3.14159\ square\ inches$$
$$\approx 28.27\ in^2$$

Look at the grid. Is $28\ in^2$ a reasonable estimate of the area of the circle?

Volume

Volume is three-dimensional and is measured in cubic units. The formula to calculate the volume of a box is:

$$V = L \times W \times H$$

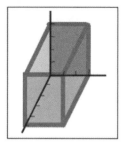

If the graph at the right is in inches, then this shape is 5 inches long, 3 inches wide, and 4 inches high.

$$V = (5\ inches) \times (3\ inches) \times (4\ inches)$$

$$V = 60\ inches^3\ or\ 60\ cubic\ inches$$

Resource **Mean, Mode, and Median**

> The **sum** of a set of numbers is the result obtained by adding the values in the set.
>
> The **size** of a set of numbers is *the number of numbers* in the set and is often designated as "**n**."

Example:

The set *A* represents the amounts Joey spent when she used her debit card yesterday. How many transactions did Joey have, and what is the total amount she spent?

$A = \{38, 14, 12, 26\}$

The size of the set is **n** = 4   Joey had 4 transactions.

How did you find **n**? Simply count the number of data values!

The sum of the set is 38 + 14 + 12 + 26 = 90.   Joey spent $90.

How did you find this? You may have used your calculator. However, you could challenge yourself to work mentally. Sometimes regrouping the numbers makes them easier to work with. Look again at the set *A*. Do you see a way to rearrange the set to make pairs of numbers you could add together more easily in your head? (Check the bottom of this page if you are stuck.)

You can regroup the numbers in your head as follows:

$$= 38 + 14 + 12 + 26$$
$$= (38 + 12) + (14 + 26)$$
$$= 50 \quad + \quad 40 =$$
$$90$$

Averages

People often talk about "averages," and you probably have an idea of what is meant by that. Now you will look at more formal mathematical ways of defining averages. In mathematics, you call an average a **measure of center** because an average is a way of measuring or *quantifying* the center of a set of data. There are different measures of center because there are different ways to define the center.

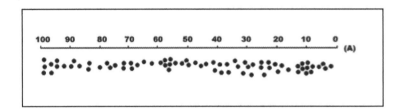

Think about a long line of people waiting to buy tickets for a concert. (Figure A shows a line about 100 feet long, and each dot represents a person in the line.) In some sections of the line, people are grouped together very closely, while in other sections of the line, people are spread out. How would you describe where the center of the line is?

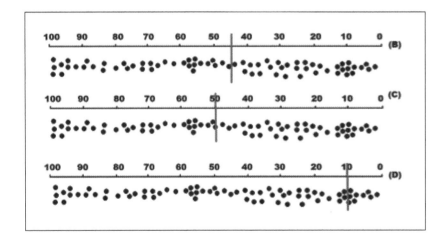

- Would you define the center of the line by finding the point at which half the people in the line are on one side and half are on the other (see Figure B)?

- Is the center based on the length of the line even though there would be more people on one side of the center than on the other (see Figure C)?

- Would you place the center among the largest groups of people (see Figure D)?

The answer would depend on what you needed the center for. When working with data, you need different measures for different purposes.

Different types of averages are described on the next page.

Mean (Arithmetic Average)

Find the average of numeric values by finding the sum of the values and dividing the sum by the number of values. The mean is what most people mean when they say "average."

Example:

$$\text{Mean} = \frac{\text{sum}}{n}$$

$$= \frac{X1 + X2 + X3 + X4 + X5}{n}$$

$$= \frac{32 + 12 + 8 + 42 + 100}{5}$$

$$= \frac{194}{5}$$

$$= 38.8$$

Manny's Purchases

	W	X
1	Shirt	32
2	Hat	12
3	Lunch	8
4	Gas	42
5	Gift Card	100

So, there are three ways to talk about this value:

1) The set is "centered" at 38.8.

2) The mean of the numbers is 38.8.

3) Manny spent an average of $38.80 per purchase.

Mode

Find the mode by finding the number(s) that occur(s) most frequently. There may be more than one mode.

Example 1:

Find the mode of 18, 23, 45, 18, 36.

The number 18 occurs twice, more than any other number, so the mode is 18.

Example 2:

Find the mode of the quiz grades. 70, 75, 75, 75, 80, 80, 85, 85, 85, 90, 95, 95

The number 75 occurs three times, as does 85. This is more than any other number, so there are two modes, 75 and 85.

Median

Find the median of a set of numbers by first arranging the data in order of size.

1) If there is an odd number of values, the median is the middle number.

2) If there is an even number of values, the median is the mean of the two middle numbers.

Example *(data set with odd number of values)*

To find the median of Manny's purchases, write the numbers in order: 8, 12, 32, 42, 100

There are 5 values (an odd number), so the median is the number in the middle.

$$8, \ 12, \ \boxed{32}, \ 42, \ 100$$
The median is 32.

Example *(data set with even number of values)*

To find the median of Joey's purchases, write the numbers in order: 12, 14, 26, 38

With an even number of values, there is no one middle number. Find the median by finding **the mean of the two middle numbers**:

$$\text{Median} = \frac{14 + 26}{2} = \frac{40}{2} = 20$$

Resource **Multiplying and Dividing Fractions**

Quantitative reasoning skill: Multiplying fractions

Another way to think about fractions is in terms of area. Look at the rectangle below. The fraction $\frac{2}{3}$ can be represented by the dark gray area found by dividing the rectangle into thirds horizontally and shading 2 of the sections as shown.

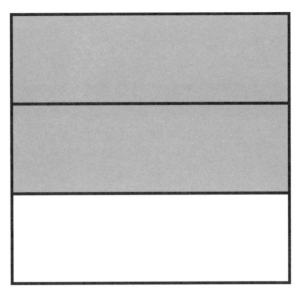

You can also think about **multiplying fractions** in terms of area of a rectangle.

Shade $\frac{2}{3}$ as indicated above. Now represent $\frac{4}{5}$ by dividing the rectangle vertically into fifths and shading 4 of the 5 sections.

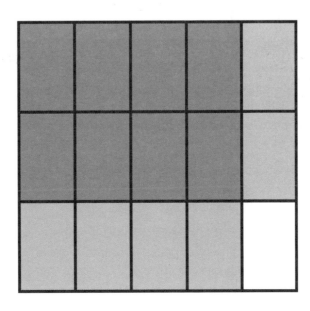

- The product $\frac{2}{3} \cdot \frac{4}{5}$ can be represented by the region that was shaded twice.

- Notice that the rectangle is now divided into 15 regions (15 = 3 × 5), and the number of those regions that are dark gray is 8 (8 = 2 × 4).

- So 8 out of 15 pieces are dark gray, or 8/15.

- This prompts a rule for multiplying fractions: multiply the numerators (2 · 4) and multiply the denominators (3 · 5), and simplify if possible.

- Therefore, $\frac{2}{3} \cdot \frac{4}{5} = \frac{8}{15}$.

Quantitative reasoning skill: Simplifying before multiplying fractions

The fact that common factors in the denominator and numerator of a number can be divided to make 1 can be used to make multiplying fractions easier. Consider the following multiplication problem.

$$\frac{2}{3} \cdot \frac{7}{8} \rightarrow \frac{14}{24} \rightarrow \frac{7 \cdot 2}{12 \cdot 2} \rightarrow \frac{7}{12}$$

If you see that there is a common factor of 2 in the numerator and denominator before multiplying, you can divide the common factors first. This makes the multiplication easier because you have smaller numbers to work with, and the simplification is complete.

$$\frac{2}{3} \cdot \frac{7}{8} \rightarrow \frac{2}{3} \cdot \frac{7}{4 \cdot 2} \rightarrow \frac{\not{2}}{3} \cdot \frac{7}{4 \cdot \not{2}} \rightarrow \frac{1 \cdot 7}{3 \cdot 4} \rightarrow \frac{7}{12}$$

This is an important concept when working with ratios with units. You will learn more about this in the Resource **Dimensional Analysis**.

Quantitative reasoning skill: Dividing fractions

Many people struggle with dividing fractions because it is difficult to visualize. A full explanation of the mathematics behind dividing fractions is beyond what the authors can do in these materials. Instead, the authors are providing you with a context that might help you remember how to divide fractions.

Suppose you have $48 to spend on going to the movies during a month. How many tickets can you buy in a month? A movie ticket costs $8. One way to think about this is that you want to know how many groups of $8 there are in $48, or 48 ÷ 8.

In the same way, suppose you had $10 to spend on downloading songs for a 1/2 dollar. (For the sake of the mathematics, you are going to express "a half of a dollar" as a fraction instead of as a decimal.) This means you want to know how many 1/2 dollars there are in $10. Your common sense probably tells you that the answer is 20 because every 1 dollar has two halves. So you multiplied 10 × 2. Look at this written as a calculation:

$$10 \div \frac{1}{2} \text{ is the same as } 10 \cdot \frac{2}{1}$$

$\frac{1}{2}$ and $\frac{2}{1}$ are called *reciprocals*.

So you say that division is the same as multiplying by the reciprocal. Here are more examples:

$$4 \div \frac{1}{2} \rightarrow 4 \cdot \frac{2}{1} \rightarrow 8$$

$$12 \div \frac{2}{3} \rightarrow 12 \cdot \frac{3}{2} \rightarrow \frac{36}{2} \rightarrow 18$$

$$\frac{4}{5} \div 2 \rightarrow \frac{4}{5} \cdot \frac{5}{2} \rightarrow \frac{20}{10} \rightarrow 2$$

Resource **Number-Word Combinations**

Combining our large number work with our rounding work can make large numbers much easier to work with in certain problems where exact values are not needed. We do this by approximating large numbers with a **number-word combination**. Consider the following examples:

25,145,561 can be rounded to the nearest million as 25,000,000

25,000,000 = 25 × 1,000,000 = **25 million**

(this is a number-word combination)

Doing that "loses" 145,561, which is quite a bit! But sometimes it doesn't matter. This number represents the population of Texas from the 2010 census. To say that the Texas population in 2010 was about 25 million people is probably good enough for most situations (source: U.S. Census Bureau, http://quickfacts.census.gov/qfd/index.html).

Option 2: round 25,145,561 to the nearest hundred-thousand, which is 25,100,000.

25,100,000 = 25.1 × 1,000,000 = **25.1 million**

This option only "lost" 45,561.

Here is another example: 1,452,900,812 rounds to 1,500,000,000 or 1.5 billion

Resource **Order of Operations**

The order of operations defines the order in which operations are performed.

General Rule	Example
1) Operations within grouping symbols, innermost first. Grouping symbols include: • Parentheses ( ) • Brackets [ ] • Fraction Bar $\frac{\square}{\square}$	$15+[12-(3+2)]-2\times3^{2}\div6$ $15+[12-(5)]-2\times3^{2}\div6$ $15+[7]-2\times3^{2}\div6$
2) Exponents	$15+[7]-2\times9\div6$
3) Multiplication and division, left to right	$15+[7]-2\times18\div6$ $15+[7]-3$
4) Addition and subtraction, left to right	$22-3$ 19

Resource **Probability, Chance, Likelihood, and Odds**

Probability

In Lesson 8, we use the word **risk** when talking about how likely it is that someone will get a disease. There are many other words that are used to describe this type of data. In mathematics, this is called a **probability**. The formula for calculating a probability is shown below. Note that you used the same type of division to calculate your percentages ("risk") in the lesson.

$$\text{Probability of an event} = \frac{\text{Number of times the event occurs}}{\text{Number of times the event could occur}}$$

If we call our event "E" and use "P" to represent probability, you sometimes will see this written as:

$$P(E) = \text{which means "the probability that E occurs"}$$

The parentheses are used differently here than their usual use as grouping symbols in mathematics; we could always use the complete phrase "the probability that E occurs." The "P(E)" notation is a shorthand for that phrase.

For example, we calculated the probability of any woman being diagnosed with lung cancer in a year:

$$\frac{\text{Number of women who get lung cancer}}{\text{Number of women}} = \frac{110,110\,\text{women}}{116,289,249\,\text{women}} = .0009469$$

So, P(any woman getting lung cancer) = .0009469

We read this statement as "the probability of any woman getting lung cancer is . . ."
Do you think that this probability is the same for all women and for each woman during their lives? You're right—this probability is an overall probability, and the probability for a given woman changes over time.

Other words that are used in describing probability are **chance** and **likelihood**.

You can look at probability as a slight change in perspective from our table in Lesson 8.A. The table showed us that 90 women in 100,000 will get lung cancer.

Probability lets us look at <u>an average woman</u>, and ask,
"What is the likelihood that <u>she</u> will get lung cancer?"

(Note: This is a general statement. Bear in mind that an individual woman can take steps to reduce her chance of lung cancer.)

Odds

In the media, you often hear **probability** and **odds** used interchangeably. However, they are <u>not</u> the same thing! Odds are best stated as a ratio.

$$\text{Odds of an event} = \frac{\text{Number of times the event occurs}}{\text{Number of times the event does not occur}}$$

So, the odds of a woman getting lung cancer =

$$\frac{\text{Number of women who get lung cancer}}{\text{Number of women who didn't get it}} = \frac{110,110 \text{ women}}{116,179,139 \text{ women}} = .0009478 : 1$$

(Do you see where the denominator came from?) We usually write odds as a ratio, such as the .0009478 to 1 in this example. The <u>odds</u> of getting lung cancer are not very different from the <u>probability</u> of getting lung cancer. Let's look at a better example:

If we flip a coin, what is the probability of getting "heads"?

$$P(\text{head}) = \frac{\text{number of heads}}{\text{number of outcomes}} = \frac{1}{2}$$

But what are the *odds* of getting "heads"?

$$\text{Odds of heads} = \frac{\text{number of heads}}{\text{number of "not heads"}} = \frac{1}{1}$$

We say "the odds of getting heads are 1 to 1."

Now say you have a jar with 5 marbles and 3 are green. You draw one marble from the jar.

The <u>probability</u> of drawing a green marble is $\frac{3}{5}$ or 60%.

The <u>odds</u> of drawing a green marble are 3 to 2.

Probabilities must be between 0 and 1. However, odds can be greater than 1, as in the example with these marbles. Probabilities can be given as percents, decimals, or ratios; odds are best given as ratios.

Resource **Properties**

On these pages, several important mathematical rules and relationships are given that can help you perform calculations in different ways. The formal names for the rules are also given. You do not have to memorize these names for <u>this</u> course, but you may use them in other math classes. If you want more help with any of the rules, use the formal names to find resources on the Internet.

The role of variables

The mathematical rules are defined in terms of **variables**. The variables are symbols, usually letters that represent numbers. You use variables to show that the rule can apply to a lot of different numbers. This is called **generalizing** because it shows that a rule can be used **in general** and not just in specific cases. The rules will be shown using variables and then give an example that uses numbers.

While mathematical rules are very important, in this course, the authors emphasize reasoning over memorizing the names of rules. As you review the rules, try to make sense of the rules so that they will become a part of your thinking.

> **Caution:**
>
> **Be sure you notice which operations**
>
> **can be used with each property!**

Communicative property

The order of addition and multiplication can be changed. It is important to remember that the Commutative Property does not apply to subtraction and division.

General Rule	Example
$a + b = b + a$	$8 + 3 = 3 + 8$
$a \times b = b \times a$	$5 \times 6 = 6 \times 5$

Distributive property

General Rule	Example
$a(b + c) = a \times b + a \times c$ also shown as $a(b + c) = ab + ac$ **Note about subtraction**: Subtraction is related to addition. The Distributive Property is shown using addition, but it also works with subtraction: $8(5 - 1) = 8 \times 5 - 8 \times 1$ **Notation**: The operation of multiplication is shown in many ways. In the above example, we see the multiplication symbol (x). We also see a number or variable in front of the parenthesis with no other symbol. For example: $6(2) = 6 \times 2$ $a(b) = a \times b$ You will learn other symbols for multiplication later in the course.	$4(3 + 1) = 4 \times 3 + 4 \times 1$ To demonstrate that these two calculations are equivalent, each side is done separately. **Left side**: Using Order of Operations, the operation inside the parentheses is done first. $4(3 + 1)$ $4(4)$ 16 **Right side**: Using the Distributive Property, the multiplication is *distributed* over the addition. $4(3 + 1)$ $4 \times 3 + 4 \times 1$ Order of Operations tells you to multiply first. $12 + 4$ 16

Division

Division is the same as multiplication by the reciprocal. You get the reciprocal of a number when you write the number as a fraction and reverse the numerator (the top number) and the denominator (bottom number). Example: The reciprocal of $\frac{2}{3}$ is $\frac{3}{2}$.

General Rule	Example
$a \div b = a \times \dfrac{1}{b}$	$15 \div 5 = 15 \times \dfrac{1}{5}$
$a \div \dfrac{b}{c} = a \times \dfrac{c}{b}$	$10 \div \dfrac{3}{5} = 10 \times \dfrac{5}{3}$
Caution:	
$15 \times \dfrac{1}{5}$ is the same as $\dfrac{1}{5} \times 15$ but $15 \div 5$ is <u>not</u> the same as $5 \div 15$	

Resource **Ratios and Fractions**

Language of ratios and fractions

A **ratio** is a comparison of two numbers by division. Ratios can be written.

- In words, such as *the ratio of males to females in class is 2 to 4.*
  This means there are two males for every four females in the class.
  Sometimes this is shown with a colon *males:females* and *2:4.*

- As a fraction, $\dfrac{\text{males}}{\text{females}} = \dfrac{2}{4}$.

Notice that the males are in the top of the fraction (the numerator) because they were mentioned first in the comparison. The females are in the bottom of the fraction (the denominator) because they were mentioned second.

$$\frac{\text{males}}{\text{females}} = \frac{2}{4} \text{ simplifies to } \frac{1}{2}. \text{ Be careful!}$$

This does <u>not</u> mean that half the class is male! It means that there is one male for every two females. What is the ratio of males to the entire class?

Think about the class as groups of 2 males and 4 females. This means that two out of every six people are male.

$$\frac{\text{males}}{\text{total}} = \frac{2}{6} = \frac{1}{3} \text{ indicates that the class is } \frac{1}{3} \text{ male.}$$

This example shows how important it is to include the context along with the math (Writing Principle #2).

What is the ratio of females to males? What fraction of the class is female? What percent of the class is female?

The ratio of females to males is $\dfrac{\text{females}}{\text{males}} = \dfrac{4}{2} = \dfrac{2}{1}$.

The ratio of females to total people is $\dfrac{\text{females}}{\text{total}} = \dfrac{4}{6} = \dfrac{2}{3} \approx 0.667 \approx 66.7\%$.

2/3 of the class is female, which is about 66.7%. More about percentages later.

Resource **Review Yourself for Exam 1**

During any class, it is important to frequently and accurately assess what you do and do not know. It is especially important before a quiz or test or when ending a unit or chapter.

Math is different from many subjects. In math, you often have to show you can complete a problem, not just remember facts or choose the right answer. Every math student has had the experience of looking at work they had previously completed or examples done in the book and thinking, "I know how to do that"—only to get home or go into a test and not be able to do a similar problem.

To check your understanding accurately, you must do problems that represent the concepts and skills you need to know. If you take time to accurately assess what you know, you can cut down on your study time. You can dedicate your study time to learning only the concepts and skills you need to understand better.

Assessing your understanding

The table on the following page lists the Exam 1 concepts and skills you should understand. This exercise helps you assess what you understand. After completing it, you will be able to prioritize your review time more effectively.

1) Assess your understanding.

- Go through the topics list and locate each concept or skill in the Lesson 1.A through 4.D Student Pages, Resource, and Assignment materials.

- If you have not used the skill in a while, do two or more problems to check your understanding.

- If you have recently used the skill and feel confident that you did it correctly, rate your understanding a 4 or 5.

- If you remember the topic but could use more practice, rate your understanding a 3.

- If you cannot remember that skill or concept, rate your understanding a 1 or 2.

Now that you have done an initial rating of your understanding, it is time to begin reviewing. Complete the remaining steps. *The goal is to have a confidence rating of 4 or 5 on all the topics in the table when you have finished your review.*

2) Start at the beginning of the set of lesson to be covered on the exam and reread the material in the lessons, the Resources, the Assignments, and your notes on the skills and concepts you rated 3 or below. Remember the Resources pages often have videos that can help.

3) Select a few problems to do. *Do not* look at the answer or your previous work for help.

4) Once you have finished the problems, check your answers. If you are not sure if you have done the problems correctly, check with your instructor, other classmates, and your previous work, or work with a tutor in the learning center.

5) Rate your confidence on this skill again. If you understand the concept better, rate yourself higher. Begin a list of topics that you want to review more thoroughly.

6) If you have time, do one or two problems on skills or concepts you rated 4 or above.

7) For topics that you need to review more thoroughly, make a plan for getting additional assistance by studying with classmates, visiting your instructor during office hours, working with a tutor in the learning center, or looking up additional information on the internet.

Lesson 1.A through 4.D

Summary of Concepts and Skills

Lesson 1.A through 4.D Concept or Skill	Rating
Working with and understanding large numbers	
Place value, naming large numbers (1.A, 2.A, 2.C, 2.D, 3.A)	
Scientific notation (2.C, 2.D)	
Calculations with large numbers (2.D, 2.D, 3.A)	
Relative magnitude and comparison of numbers (2.C, 2.D)	
Estimation and calculation	
Rounding (3.A)	
Fractions and decimals (3.C, 3.D)	
Relationship of multiplication and division (3.C, 4.A, 4.B)	
Order of operations (4.A, 4.B)	
Percentages and ratios	
Estimations with fraction and percent benchmarks (3.C, 4.A, 4.B)	
Calculate percentages (3.D, 4.A, 4.B)	
Write and understand ratios (2.C, 2.D)	
Read and interpret graphs	
Read line graphs and determine absolute and relative change (4.C)	
Read pie charts (circle graphs) and calculate quantities given the base value (4.D)	

Resource **Review Yourself for Exam 2**

As with Lesson 1.A through 4.D, you should assess your understanding of Lesson 5.A through 9.D to prepare for Exam 2. Your instructor may give you specific assignments for your review in addition to this self-assessment. Completing this self-assessment first can help you to determine areas where you need additional help prior to beginning any additional assignment.

Mathematics tends to be a cumulative subject. As such, you also need to be sure you are still comfortable with the "toolkit" topics from Lesson 1.A through 4.D. These skills are the foundation of many of the concepts in Lesson 5.A through 9.D and reviewing them will make your Exam 2 review more productive. In addition, Exam 2 may include some questions that are specifically tied to Exam 1.

Assessing your understanding

The table on the following pages lists the Lesson 5.A through 9.D concepts and skills you should understand. This exercise helps you assess what you understand. After completing it, you will be able to prioritize your review time more effectively.

1. Assess your understanding.

Go through the topics list. Ask yourself if you understand the description of the topic. Can you picture the type of problem? For example, the first set of topics is titled "Graphical Displays of Data." You should be able to picture histograms and dotplots of different shapes. You should also know the difference between a histogram and a relative histogram. Consider making up quick examples of these graphical displays. If you can do that, you are set for that topic! In addition, you could…

Locate each concept or skill in the Lesson 5.A through 9.D Student Pages and Assignment materials.

If you have not used the skill in a while, do two or more problems to check your understanding.

If you have recently used the skill and feel confident that you did it correctly, rate your understanding a 4 or 5.

If you remember the topic but could use more practice, rate your understanding a 3.

If you cannot remember that skill or concept, rate your understanding a 1 or 2.

Now that you have done an initial rating of your understanding, it is time to begin reviewing. Complete the remaining steps. *The goal is to have a confidence rating of 4 or 5 on all the topics in the table when you have finished your review of the second set of lessons.*

2. Start at the beginning of the set of lesson to be covered on the exam and reread the material in the lessons, the Resources, the Assignments, and your notes on the skills and

concepts you rated 3 or below. Check the Resources Table of Contents to see if there is a resource page associated with the skill.

3. Select a few problems to do. *Do not* look at the answer or your previous work to help you.

4. Once you have finished the problems, check your answers. If you are not sure if you have done the problems correctly, check with your instructor, other classmates, and your previous work or work with a tutor in the learning center.

5. Rate your confidence on this skill again. If you understand the concept better, rate yourself higher. Begin a list of topics that you want to review more thoroughly.

6. If you have time, do one or two problems on skills or concepts you rated 4 or above.

7. For topics that you need to review more thoroughly, make a plan for getting additional assistance by studying with classmates, visiting your instructor during office hours, working with a tutor in the learning center, or looking up additional information on the internet.

Lesson 5.A through 9.D

Summary of Concepts and Skills

Lesson 5.A through 9.D Concept or Skill	Rating
Graphical Displays of Data	
Interpret stem-and-leaf plots, split-stems, and back-to-back plots (5.A)	
Interpret frequency, relative, and cumulative frequency tables (5.B)	
Read and interpret histograms and relative histograms (5.C)	
Read and interpret dotplots (5.D)	
Analyze the shapes of distributions (5.C, 5.D)	
Numerical Summaries of Data	
Calculate and interpret mean, mode, median (5.D, 6.A, 6.C)	
Create a data set to meet criteria (6.A)	
Make and justify decisions based on data (6.C)	
Determine the five-number summary (6.D)	
Create and interpret boxplots (6.D)	
Application of Instructions	
Order of operations (7.A through 7.D)	
Properties that allow flexibility in calculations: Distributive Property, Commutative Property (7.A, 7.D)	
Properties that allow flexibility in calculations: Distributive Property, Commutative Property (7.A through 7.D)	
Perform multi-step calculations (7.A through 7.D)	
Writing expressions and spreadsheet formulas (7.A through 7.D)	

Interpreting Categorical Data	
Comparing data expressed in different forms (8.A, 8.B)	
Identify missing information (8.A)	
Calculate absolute and relative change (8.C, 8.D)	
Calculate and interpret marginal and conditional probabilities (9.A and 9.B)	
Analyzing accuracy of medical test results (9.C, 9.D)	

8. Now, if needed, repeat the process for the material from Lesson 1.A through 4.D. Many of the concepts in Lesson 5.A through 9.D are extensions or refinements of concepts from Lesson 1.A through 4.D. Specific skills that did not see as much use during Lesson 5.A through 9.D are given.

Lessons 1.A through 4.D Concept or Skill	Rating
Working with and Understanding Large Numbers	
Place value, naming large numbers (1.A, 2.A, 2.C, 2.D, 3.A)	
Scientific notation (2C, 2.D)	
Calculations with large numbers (2.C, 2.D, 3.A)	
Relative magnitude and comparison of numbers (2.C, 2.D)	
Percentages and Ratios	
Estimations with fraction and percent benchmarks (3.C, 4.A, 4.B)	
Read and Interpret Graphs	
Read line graphs and determine absolute and relative change (4.C)	
Read pie charts (circle graphs) and calculate quantities given the base value (4.D)	

Resource **Review Yourself for Exam 3**

As with the previous Exams, you should assess your understanding of Lesson 10.A through 14.D to prepare for the Exam 3. Your instructor may give you specific assignments for your review in addition to this self-assessment. Check with your instructor to see if Exam 3 will include concepts from Lesson 1.A through 9.D that were not specifically needed in Lesson 10.A through 14.D. If so, complete the condensed Lesson 1.A through 9.D checklists as well.

Assessing your understanding

The table on the following pages lists the Lesson 10.A through 14.D concepts and skills you should understand. This exercise helps you assess what you understand. After completing it, you will be able to prioritize your review time more effectively.

1. Assess your understanding.

Go through the topics list. Ask yourself if you understand the description of the topic. Can you picture the type of problem? For example, the first set of topics is titled "Graphical Displays of Data." You should be able to picture histograms and dotplots of different shapes. You should also know the difference between a histogram and a relative histogram. Consider making up quick examples of these graphical displays. If you can do that, you are set for that topic! In addition, you could…

Locate each concept or skill in the Lesson 10.A through 14.D Student Pages and Assignment materials.

If you have not used the skill in a while, do two or more problems to check your understanding.

If you have recently used the skill and feel confident that you did it correctly, rate your understanding a 4 or 5.

If you remember the topic but could use more practice, rate your understanding a 3.

If you cannot remember that skill or concept, rate your understanding a 1 or 2.

Now that you have done an initial rating of your understanding, it is time to begin reviewing. Complete the remaining steps. *The goal is to have a confidence rating of 4 or 5 on all the topics in the table when you have finished your review of the third set of lessons.*

2. Start at the beginning of the set of lesson to be covered on the exam and reread the material in the lessons, the Resources, the Assignments, and your notes on the skills and concepts you rated 3 or below. Check the Resource Table of Contents to see if there is a resource page associated with the skill.

3. Select a few problems to do. *Do not* look at the answer or your previous work to help you.

4. Once you have finished the problems, check your answers. If you are not sure if you have done the problems correctly, check with your instructor, other classmates, and your previous work or work with a tutor in the learning center.

5. Rate your confidence on this skill again. If you understand the concept better, rate yourself higher. Begin a list of topics that you want to review more thoroughly.

6. If you have time, do one or two problems on skills or concepts you rated 4 or above.

7. For topics that you need to review more thoroughly, make a plan for getting additional assistance by studying with classmates, visiting your instructor during office hours, working with a tutor in the learning center, or looking up additional information on the internet.

Lesson 10.A through 14.D
Summary of Concepts and Skills

Lesson 10.A through 14.D Concept or Skill	Rating
Using Ratios	
Understand meaning of equivalent ratios in context (10.A)	
Calculate a unit rate (10.B)	
Use ratios and proportionality to calculate new values (10.B)	
Calculate and interpret absolute and relative change (10.D)	
Use units with ratios (10.A through 10.D)	
Geometric Reasoning	
Understand concepts of and units for linear measurement, area, and volume (11.A through 11.C)	
Identify and use appropriate geometric formulas (11.A through 11.C)	
Making Conversions	
Understand use of units in making conversions (12.A through 12.D)	
Use dimensional analysis to make a conversion involving multiple conversion factors (12.A through 12.D)	
Using Formulas and Algebraic Expressions	
Understand the use of variables in formulas and algebraic expressions, including the appropriate way to define a variable (13.A through 13.C)	
Understand the role of a constant in a formula (13.A through 13.C)	
Use a formula to solve for a value (13.A through 13.D)	
Creating and Solving Proportions	
Solve a linear equation in one variable (14.A through 14.D)	
Interpret the solution to an equation (14.A through 14.D)	

8. Now repeat the process for the material from Lesson 1.A through 9.D. The table on the next page lists specific skills that did not see as much use during Lesson 10.A through 14.D.

Lesson 1.A through 9.D Concept or Skill	Rating
Numerical Skills	
Scientific notation (2.C, 2.D)	
Rounding (3.A)	
Estimations with fraction and percent benchmarks (3.C, 4.A, 4.B)	
Calculate percentages (3.D, 4.A, 4.B)	
Calculate percentages from two-way tables (9.A)	
Use percentages as probabilities and ratios (9.C, 9.D)	
Graphical Displays of Data	
Read line graphs and determine absolute and relative change (4.C)	
Recognize of graphs due to different scales (4.C)	
Read and interpret histograms and relative histograms (5.C)	
Read and interpret dotplots (5.D)	
Analyze the shapes of distributions (5.C, 5.D)	
Read pie charts (circle graphs) and calculate quantities given the base value (4.D)	
Interpret stem-and-leaf plots, split-stems, back-to-back plots (5.A)	
Interpret frequency, relative, and cumulative frequency tables (5.B)	
Numerical Summaries of Data	
Calculate and interpret mean, mode, median (5.D, 6.A, 6.C)	
Create a data set to meet criteria (6.A)	
Make and justify decisions based on data (6.C)	
Determine the five-number summary (6.D)	
Create and interpret boxplots (6.D)	
Interpreting Categorical Data	
Comparing data expressed in different forms (8.A, 8.B)	
Identify missing information (8.A)	
Calculate absolute and relative change (8.C, 8.D)	
Calculate, interpret marginal and conditional probabilities (9.A, 9.B)	
Analyzing accuracy of medical test results (9.C, 9.D)	

Resource **Review Yourself for Exam 4**

It is easy to say to yourself, "I know how to create a linear equation to model data." However, you do not know if that assessment is accurate unless you have evidence showing that you are able to create a linear equation to model data. To accurately assess your understanding, you must provide evidence by doing problems and checking to see whether you did them correctly.

Assessing your understanding

The table on the following pages lists the Lesson 15.A through 19.E concepts and skills you should understand. This exercise helps you assess what you understand. After completing it, you will be able to prioritize your review time more effectively.

1. Assess your understanding.

Go through the topics list and locate each concept or skill in the indicated Student Pages, the Assignments associated with those lessons, or the Resources.

If you have not used the skill in a while, do two or more problems to check your understanding.

If you have recently used the skill and feel confident that you did it correctly, rate your understanding a 4 or 5.

If you remember the topic but could use more practice, rate that topic with "3".

If you do not remember that skill or concept, rate your understanding a 1 or 2.

Now that you have done an initial rating of your understanding, it is time to begin reviewing. Complete the remaining steps. *The goal is to have a confidence rating of 4 or 5 on all the topics in the table when you have finished your review of that set of lessons.*

2. Reread the material in the lessons, the assignments, and your notes on the skills and concepts you rated 3 or below.

3. Select a few problems to do. *Do not* look at the answer or your previous work to help you. Once you have finished the problems, check your answers. If you are not sure if you have done the problems correctly, check with your instructor, other classmates, and your previous work *or* work with a tutor in the learning center.

5. Rate your confidence on this skill again. If you understand the concept better, rate yourself higher. Begin a list of topics that you want to review more thoroughly.

6. If you have time, do one or two problems on skills or concepts you rated 4 or above.

7. For topics that you need to review more thoroughly, make a plan for getting additional assistance by studying with classmates, visiting your instructor or the tutoring center, or looking up additional information on the internet.

Lesson 15.A through 19.E
Summary of Concepts and Skills

Concept or Skill	Rating
Linear Models	
Solve an equation or formula for a variable (15.C)	
Write and solve proportions (15.A, 15.B, 15.C, 15.E)	
Solve complex equations with multiple variable terms and variables in the denominator (15.D)	
Understand the concept of slope as a unit rate (16.A, 16.B)	
Compare and contrast slopes (16.B, 17.C)	
Understand the role of units in a linear equation (16.A through 16.E)	
Calculate and interpret slope (16.A through 16.E, 17.A through 17.D)	
Create a linear equation to model data (16.D, 16.E, 17.C, 17.D)	
Identify and interpret vertical intercept, horizontal intercept, and slope from a graph. (16.D, 16.E, 17.A through 17.D, 19.D)	
Identify the slope and vertical intercept from an equation (17.A through 17.D)	
Understand that a linear model is defined by a constant rate of change (17.A, 17.B, 19.D)	
Create, interpret, use, and translate between the four representations of a linear model (verbal, table, graph, equation) (16.A through 16.E, 17.A through 17.D, 19.D)	
Identify a model as linear based on any of the four representations (17.A through 17.D, 19.D)	
Understand the limitations of models based on data (17.C, 17.D)	
Use equations, tables, and graphs to solve or estimate solutions to problems (17.A through 17.D, 19.D)	
Write and use linear equations based on a percentage increase or decrease (18.A, 18.B)	

Exponential Models	
Create, interpret, and use the four representations of an exponential model (verbal, table, graph, equation) (18.C, 18.D, 19.A through 19.C)	
Write and use an exponential model (18.C, 18.D, 19.A through 19.C)	
Understand that an exponential model is defined by a percentage change (19.A through 19.C)	
Identify an exponential model as growth or decay (19.A through 19.C)	
Compare and Contrast	
Linear and Exponential Contexts (18.D, 19.C, 19.D)	

Resource **Review Yourself for the Final Exam**

As you have seen throughout this course, it is important to assess your skills accurately, with evidence to support your assessment. It is easy to say to yourself, "I know how to create a linear equation to model data." However, you do not know if that assessment is accurate unless you have evidence showing that you are able to create a linear equation to model data. To accurately assess your understanding with evidence, you must do problems and check to see whether you did them correctly.

Assessing your understanding

At the end of each set of lessons, you were given a list of skills and concepts to help you assess your understanding and organize your review process. You will use those same skill-and-concept lists to prepare for the final exam. Your instructor may assign specific problems for you to work and give you deadlines for the different review sections.

Here is a reminder of the steps to use in reviewing for each exam.

1. Assess your understanding.

    -   Go through the topics list and locate each concept or skill in the indicated Student Pages, the Assignments associated with those lessons, or the Resources.
    -   If you have not used the skill in a while, do two or more problems to check your understanding.
    -   If you have recently used the skill and feel confident that you did it correctly, rate your understanding a 4 or 5.
    -   If you remember the topic but could use more practice, rate your understanding a 3.
    -   If you cannot remember that skill or concept, rate your understanding a 1 or 2.

    Now that you have done an initial rating of your understanding, it is time to begin reviewing. Complete the remaining steps. The goal is to have a confidence rating of 4 or 5 on all the topics in the table when you have finished your review of the set of lessons to be covered on the exam.

2. Start at the beginning of the set of lesson to be covered on the exam, and reread the material in the lessons, the assignments, and your notes on the skills and concepts you rated 3 or below.

3. Select a few problems to do. *Do not* look at the answer or your previous work to help you.

4. Once you have finished the problems, check your answers. If you are not sure if you have done the problems correctly, check with your instructor, other classmates, and your previous work *or* work with a tutor in the learning center.

5. Rate your confidence on this skill again. If you understand the concept better, rate yourself higher. Begin a list of topics that you want to review more thoroughly.

6. If you have time, do one or two problems on skills or concepts you rated 4 or above.

7. For topics that you need to review more thoroughly, make a plan for getting additional assistance by studying with classmates, visiting your instructor during office hours, working with a tutor in the learning center, or looking up additional information on the Internet.

Studying for a final exam is challenging because there is so much material to review. That is why it is important to start early and break your review up into smaller chunks of work. You should set deadlines for your initial review of each set of lessons covered on each exam using Steps 1–7. When you have your complete list for the more in-depth review, you can make a plan for accomplishing that work.

Lesson 1.A through 4.D
Summary of Concepts and Skills

Deadline for initial review:

Lesson 1.A through 4.D Concept or Skill	Rating
Working with and understanding large numbers	
Place value, naming large numbers (1.A, 2.A, 2.C, 2.D, 3.A)	
Scientific notation (2.B, 2.C, 2.D)	
Calculations with large numbers (2.D, 2.D, 3.A)	
Relative magnitude and comparison of numbers (2.C, 2.D)	
Estimation and calculation	
Rounding (3.A)	
Fractions and decimals (3.C, 3.D)	
Relationship of multiplication and division (3.C, 4.A, 4.B)	
Order of operations (4.A, 4.B)	
Percentages and ratios	
Estimations with fraction and percent benchmarks (3.C, 4.A, 4.B)	
Calculate percentages (3.D, 4.A, 4.B)	
Write and understand ratios (2.C, 2.D)	
Read and interpret graphs	
Read line graphs and determine absolute and relative change (4.C)	
Read pie charts (circle graphs) and calculate quantities given the base value (4.D)	

Lesson 5.A through 9.D
Summary of Concepts and Skills

Deadline for initial review:

Lesson 5.A through 9.D Concept or Skill	Rating
Graphical Displays of Data	
Interpret stem-and-leaf plots, split-stems, and back-to-back plots (5.A)	
Interpret frequency, relative, and cumulative frequency tables (5.B)	
Read and interpret histograms and relative histograms (5.C)	
Read and interpret dotplots (5.D)	
Analyze the shapes of distributions (5.C, 5.D)	
Numerical Summaries of Data	
Calculate and interpret mean, mode, median (5.D, 6.A, 6.C)	
Create a data set to meet criteria (6.A)	
Make and justify decisions based on data (6.C)	
Determine the five-number summary (6.D)	
Create and interpret boxplots (6.D)	
Application of Instructions	
Properties that allow flexibility in calculations: Distributive Property, Commutative Property (7.A, 7.D)	
Order of operations (7.A through 7.D)	
Properties that allow flexibility in calculations: Distributive Property, Commutative Property (7.A through 7.D)	
Perform multi-step calculations (7.A through 7.D)	
Writing expressions and spreadsheet formulas (7.A through 7.D)	
Interpreting Categorical Data	
Comparing data expressed in different forms (8.A, 8.B)	
Identify missing information (8.A)	
Calculate absolute and relative change (8.C, 8.D)	
Calculate and interpret marginal and conditional probabilities (9.A and 9.B)	
Analyzing accuracy of medical test results (9.C, 9.D)	

Lesson 10.A through 14.D
Summary of Concepts and Skills

Deadline for initial review:

Lesson 10.A through 14.D Concept or Skill	Rating
Using Ratios	
Understand meaning of equivalent ratios in context (10.A)	
Calculate a unit rate (10.B)	
Use ratios and proportionality to calculate new values (10.B)	
Calculate and interpret absolute and relative change (10.D)	
Use units with ratios (10.A through 10.D)	
Geometric Reasoning	
Understand concepts of and units for linear measurement, area, and volume (11.A through 11.C)	
Identify and use appropriate geometric formulas (11.A through 11.C)	
Making Conversions	
Understand use of units in making conversions (12.A through 12.D)	
Use dimensional analysis to make a conversion involving multiple conversion factors (12.A through 12.D)	
Using Formulas and Algebraic Expressions	
Understand the use of variables in formulas and algebraic expressions, including the appropriate way to define a variable (13.A through 13.C)	
Understand the role of a constant in a formula (13.A through 13.C)	
Use a formula to solve for a value (13.A through 13.D)	
Creating and Solving Proportions	
Solve a linear equation in one variable (14.A through 14.D)	
Interpret the solution to an equation (14.A through 14.D)	

Lesson 15.A through 19.E
Summary of Concepts and Skills

Deadline for initial review:

Concept or Skill	Rating
Linear Models	
Solve an equation or formula for a variable (15.C)	
Solve complex equations with multiple variable terms and variables in the denominator (15.D)	
Write and solve proportions (15.A, 15.B, 15.C, 15.E)	
Understand the concept of slope as a unit rate (16.A, 16.B)	
Compare and contrast slopes (16.B, 17.C)	
Understand the role of units in a linear equation (16.A through 16.E)	
Calculate and interpret slope (16.A through 16.E, 17.A through 17.D)	
Create a linear equation to model data (16.D, 16.E, 17.C, 17.D)	
Identify and interpret vertical intercept, horizontal intercept, and slope from a graph. (16.D, 16.E, 17.A through 17.D, 19.D)	
Identify the slope and vertical intercept from an equation (17.A through 17.D)	
Understand that a linear model is defined by a constant rate of change (17.A, 17.B, 19.D)	
Create, interpret, use, and translate between the four representations of a linear model (verbal, table, graph, equation) (16.A through 16.E, 17.A through 17.D, 19.D)	
Identify a model as linear based on any of the four representations (17.A through 17.D, 19.D)	
Understand the limitations of models based on data (17.C, 17.D)	
Use equations, tables, and graphs to solve or estimate solutions to problems (17.A through 17.D, 19.D)	
Write and use linear equations based on a percentage increase or decrease (18.A, 18.B)	

Exponential Models	
Create, interpret, and use the four representations of an exponential model (verbal, table, graph, equation) (18.C, 18.D, 19.A through 19.C)	
Write and use an exponential model (18.C, 18.D, 19.A through 19.C)	
Understand that an exponential model is defined by a percentage change (19.A through 19.C)	
Identify an exponential model as growth or decay (19.A through 19.C)	
Compare and Contrast Linear and Exponential Contexts (18.D, 19.C, 19.D)	

Resource **Rounding and Estimation**

Another important skill you used in this lesson is rounding. You often round numbers when you are trying to make sense out of them or make comparisons and do not need exact numbers. In this lesson, you estimated the length of a line of one billion people by measuring the people in your class as a sample. Since this was an estimate, there was no need to keep exact values in your results.

In this course, you will talk about different types of *estimation*.

- **Educated guess:** One type of estimation might be called an "educated guess" about something that has not been measured exactly. In an upcoming lesson, you will use estimations of the world population. This quantity cannot be measured exactly—it would be impossible to count how many people live on Earth at any given time. Scientists can use good data and mathematical techniques to estimate the population, but it will always be an estimate.

- **Convenient estimation:** Sometimes estimations are used when it is inconvenient or not worthwhile to make an exact count. Imagine that you need to know how much paint to buy to paint the baseboard trim in your house. (The baseboard trim is the piece of wood that follows along the bottom of the walls.) You need to know the length of the baseboard. You could measure the length of each wall to the nearest 1/8 inch and carefully subtract the width of halls and doors. It would be much quicker and just as effective to measure to the nearest foot or half foot. If you were cutting a piece of baseboard to go along the floor, however, you would want an exact measurement.

- **Estimated calculation:** This usually involves rounding numbers to make calculations simpler. A future lesson will focus on estimating and calculating percentages. In this course, you will find that percentages are used in many contexts. One of the most important skills that you will develop is understanding and being comfortable working with percentages in a variety of situations.

Resource **Scientific Notation**

Numbers can be written in many different forms. Some examples are:

$50 = 5 \times 10$

$400 = 40 \times 10$ or 4×100 or 4×10^2

$7,000 = 700 \times 10$ or 70×100 or 7×1000 or 7×10^3

Likewise: 52 can be written as 5.2×10

473 can be written as 47.3×10 or 4.73×100 or 4.73×10^2

$7,549 = 754.9 \times 10$ or 75.49×100 or 7.549×1000 or 7.549×10^3

Look at the last form in each row. What do they have in common? *They all have a single non-zero digit in front of the decimal place.* This means that we have a number greater than or equal to 1 but less than 10, multiplied by a power of 10. This is **scientific notation.**

> A number in *scientific notation* is written in the form:
>
> $a \times 10^n$ where $1 \le a < 10$; and n is an integer.
>
> 1 is included   and   10 is not included

7.549×10^3 is in scientific notation (7 is a number between 1 and 10).

75.49×10^2 is _not_ in scientific notation (because 75 is not between 1 and 10).

7.549×100^3 is _not_ in scientific notation because 100^3 is not written as a power of 10.

Resource **Slope**

Calculating Slope

In Lesson 16.C, you developed the formula for slope. In a linear relationship containing points (x_1, y_1) and (x_2, y_2), slope can be found in the following manner:

$$slope = m = \frac{y_2 - y_1}{x_2 - x_1}$$

Keep in mind:

- Slope can be found using any two points on a line. They can be given in a problem description, presented in a table, or read from a graph of the line.

- The "y" values are always the dependent values in the problem, even if you choose a different variable to represent them.

- The "x" values are always the independent values in the problem, even if you choose a different variable to represent them.

The points in each of the following examples are presented in different ways. Review the example, then try your own.

1) Using two points to find slope.

Example:

$(3, -2)$ **and** $(4, 7)$

$$slope = m = \frac{y_2 - y_1}{x_2 - x_1}$$

$$m = \frac{7 - (-2)}{4 - 3} = \frac{9}{1} = 9$$

You try:

$(-3, -5)$ **and** $(4, 16)$

2) Use values in a table to calculate slope.

Example:

Time (sec)	5	10	15	20
Distance (yd)	40	75	110	145

Choose any two points: $(5, 40)$ and $(15, 110)$

$$slope = m = \frac{y_2 - y_1}{x_2 - x_1}$$ $$m = \frac{110\,\text{yd} - 40\,\text{yd}}{15\,\text{sec} - 5\,\text{sec}} = \frac{70\,\text{yd}}{10\,\text{sec}} = \frac{7\,\text{yd}}{1\,\text{sec}}$$

We interpret the slope with the statement: *"The distance changes seven yards every one second."*

You try:

Snacks	5	10	15	20
Cost ($)	15	30	45	60

3) Use values from a graph to calculate slope.

Example: **You try:**

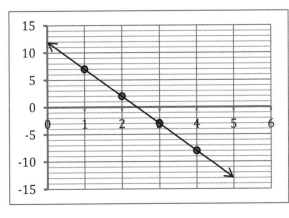   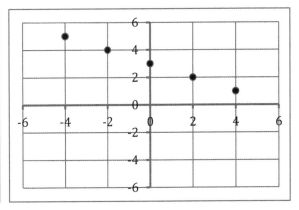

Select any two points from the graph.

$(2, 2)$ and $(3, -3)$

$$slope = m = \frac{y_2 - y_1}{x_2 - x_1}$$

$$m = \frac{-3 - 2}{3 - 2} = \frac{-5}{1} = -5$$

Resource **Understanding Visual Displays of Information**

Asking questions about displays

Data are increasingly presented in a variety of forms intended to interest you and to invite you to think about the importance of these data and how they might affect your lives.

Some common displays are:

- pie charts
- scatterplots
- histograms and bar graphs
- line graphs
- tables
- pictographs

What questions should you ask yourself when you study a visual display of information?

- What is the title of the chart or graph?
- What question is the data supposed to answer? (For example: How many males versus females exercise daily?)
- How are the columns and rows labeled? How are the vertical and horizontal axes labeled?
- Select one number or data point and ask, "What does this mean?"

The chart on the next page can help you understand what some basic types of visual displays of information tell you and what questions they usually answer.

This looks like a ...	This visual display is usually used to ...	For example, it can be used to show ...
Pie chart	Show the relationships between different parts compared to a whole.	How time is used in a 24-hour cycle. How money is distributed. How something is divided up.
Line graph	Show trends over time. Compare trends of two different items or measurements.	What seems to be increasing. What is decreasing. How the cost of gas has increased in the last 10 years. Which of these foods (milk, steak, cookies, eggs) has risen most rapidly in price compared to the others.
Histogram or bar graph	Compare data in different categories. Show changes over time.	How a population is broken up into different age categories. How college tuition rates are changing over time.
Table	Organize data to make specific values easy to read. Break data up into overlapping categories.	The inflation rates over a period of years. How a population is broken into males and females of different age categories.

Resource **Writing Principles**

Writing background

You might be surprised that you are asked to write short responses to questions in this class. Writing in a math class? This course emphasizes writing for the following two reasons:

- Writing is a learning tool. Explaining things such as the meaning of data, how you calculated the data, or how you know your answer is correct deepens your own understanding of the material.

- Communication is an important skill in quantitative literacy. Quantitative information is used widely in today's world in products such as reports, news articles, publicity materials, advertising, and grant applications.

Understanding the task

One important strategy in writing is to make sure you understand the task. In this course, your tasks will be questions in assignments, but in other situations, the task might be a question on a report form, instructions from your employer, or a goal that you set for yourself. To begin to write successfully, ask yourself the following questions:

- What is the topic of the writing task?

- What is the task telling me to do? Some examples are given below:

  o Describe how you found the answer.

  o Explain why you think you have the right answer.

  o Reflect on the process of coming up with the answer.

  o Make a prediction about the next data point.

  o Compare two data points or the answers to two parts of the problem.

- What information am I given to help me with the task?

Look at this example and the answers to these questions.

> In Preview Assignment 2.A, you read about monitoring your readiness. Explain briefly why it is important to monitor your readiness before coming to class.

- What is the topic of the writing task? [Answer: Monitoring whether I am ready for the next class meeting.]

- What is the task telling me to do? [Answer: It is asking me to *explain* why "monitoring readiness" is important.]

- What information am I given to help me with the task? [Answer: I can look back at the Preview Assignment for 2.A if I need to refresh on this topic.]

Writing principles

Principle #1   If the problem has words, so should the answer!

(Have you noticed almost all of the problems in this class have words?)

Strive to be neat and to use proper grammar, spelling, and punctuation.

Principle #2   Each answer should be in a complete sentence that stands on its own, which means that the relevant information from the problem should be in the answer. **The readers should understand what you are trying to say even if they have not read the question or writing prompt.** Relevant information includes:

- Information about context.

- Quantitative information.

Refer back to question 8 from Student Page 2.A:

What are some factors you think may have led to this change in doubling times?

Insufficient response: *Improved health care, better food.*

Good response: *The world's population has increased rapidly. This increase may be due to factors such as improved health care, better food supplies, and clean water.*

Principle #3   If you use tables or graphs in your response, be sure they are clearly and thoroughly labeled.

Principle #4   Let the reader know if you are making any assumptions. One example of this could be when there is unclear information in the problem.

Mary's bedroom is 10 feet wide, 12 feet long, and 9 feet high. If a gallon of paint covers 300 square feet, how many gallons should Mary buy?

After doing the math and writing your answer, you might say, *"I didn't know how many windows and doors Mary has, so I just didn't take them into account in my calculations."*